Hope in the Soil

A Topical Compilation of the Writings of Ellen G. White on Farming and Gardening—and the Role of Agriculture in Character Development, Evangelism, and Preparing a People for the Second Coming

Ellen G. White

Compiled by Cari Haus

Copyright © 2015 Cari Haus

P. O. Box 7
Cedar Lake, Michigan

The compiler assumes full responsibility for the accuracy and interpretation of the Ellen White quotations cited in this book.

Table of Contents

Preface and Acknowledgements 7

The Relation of True Education to the Farm (an article by Ellen G. White) 9

Chapter #1: The Key to Unlocking God's Word 13
    Part 1: Nature the Most Effective Way to Reach Minds Far from God 17
    Part 2: A Warning from Sodom 21
    Part 3: Farming a Consistent Way to Reach Souls through Nature 24

Chapter #2: The Farm as God's Training School 27
    Part 1: Youth in Critical Spiritual Need 27
    Part 2: How the Youth Can Be Reached 29
    Part 3: Education Close to Nature Superior for Spiritual Growth 31
    Part 3: Nature Especially Valuable in Teaching Small Children 32
    Part 4: Farming Invaluable in Training Evangelists 35
    Part 5: God's Ownership of School Farms 40

Chapter #3: Godly Farms—Schools and Family-owned—as Beacons of Light . 43
    Part 1: God's Original Homesteading Plan 43
    — (1): A Plan Developed Before the World Was 43
    — (2): Advantages of Country Living for Child-rearing 49
    — (3): Dangers to Children from City Living 51
    — (4): Saving Enough for a Home 52

— (5): Prudence and Wisdom Needed—Rash Moves Not to Be Made 53
Part 2: Powerful Testimony of Beautiful, Well-kept Grounds and Farms 54
Part 3: Farmers Highly Effective as Missionaries 57
Part 4: God's Economic Plan for Helping the Poor 61
Part 5: Ellen White's Missionary Work Helping Farm Families Get Started . 64
Part 6: Scattered as Beacons of Light 67

Chapter #4: Blessings to be Obtained through Involvement in Farming 75
Blessing #1: Character Traits Developed 75
— (1): Faithfulness in Little Things 75
— (2): Cooperation with God 79
— (3): Grace and Spiritual Nerve 80
— (4): Industry, Religious Thrift, and Dependence on God 80
— (5): Intelligence and Innovation 81
— (6): Leadership Ability 84
— (7): Perseverance, Discipline, and Patience 86
— (8): Promptness and Responsibility 87
— (9): Self-Sufficiency 88
— (10): Speed and Efficiency 93
— (11): Thoroughness and Diligence 94
Blessing #2: Unseen Advantages and Heaven-planned Partnerships 96
Blessing #3: Physical Benefits to Students and Others 97
Blessing #4: Spiritual Benefits to Students and Others 104
Blessing #5: Improved Thinking Ability and Vigor of Thought 107
Blessing #6: Financial Success 108

— Financial Blessing #1: Bountiful Harvests through Diligent Farming 108
— Financial Blessing #2: Good Yields from Unpromising Ground 111
— Financial Blessing #3: Self-sufficiency during the Sunday Law Crisis ... 114

Chapter #5: Characteristics of Model Schools that Follow God's Plan 115

Characteristic #1: Nature and the Bible as Primary Textbooks 115

Characteristic #2: Hand-in-Hand Integration with an Agricultural Program 121

Characteristic #3: Balance between Mental and Physical Labor 127

— (1): Manual Labor an Integral Part of Curriculum 130

— (2): Hard Work Employed as a Safeguard against Sin 137

Characteristic #4: Rural or Country Location 142

— (1): Plenty of Land Reserved for Use by the School 148

Characteristic #5: Every Student Learning a Practical Trade 154

Characteristic #6: Farming Supercedes Love of Sports and Amusements 157

Characteristic #7: Teachers and Ministers Working Alongside Students 161

(1) A Competent Farm Manager to be Employed 166

Characteristic #8: Self-sufficiency 166

— (1): Farming a Good Investment for Schools 166

— (2): Financial Losses during the Start-up Phase 167

— (3): The School to Grow its Own Food 168

— (4): Farming a Means of Financial Support for the School 170

— (5): Why Some Farms Do Poorly 173

Characteristic #9: Obedience to God's Plan Vital to Farming Success 175

Characteristic #10: Small Schools—Connected with Small Sanitariums 180

Chapter #6: a School Farms that Followed God's Plan and Prospered 182
   God's Farm at Madison College 182
   God's Farm at Avondale College 184
   — (1) A Letter from Avondale 187

Chapter #7: No Time to Lose 191
   (1): Promises of Help and Success 195
   (2): Results of Failure to Obey 198

Preface and Acknowledgements

"There is hope in the soil, but brain and heart and strength must be brought into the work of tilling it." (Special Testimonies on Education p. 100)

"The hope of advancing the cause of God in this country is in creating a new moral taste in love of work, which will transform mind and character." (Fundamentals of Christian Education p. 322)

This compilation and syllabus is the result of many hours of study on the writings of Ellen G. White with regard to farming and agriculture. This work is not something I started or did by myself. While I did research various topics and added numerous quotes, many of the passages included in this compilation were drawn from the works of others. Specifically, I would like to mention and thank:

David Obermiller, Farm Manager at *Harvest Fields Organic Farm* at Fresno Adventist Academy (California). David is in the process of publishing two books on farming:

Food Over Faith? The Prophetic Relevance of Farming in Adventism

and

Counsels on Gardening: Ending Where it Began

David's first book, Food over Faith, offers an incredible overview of God's use of farming as a training tool from Old Testament times to the present. I had the great privilege of editing that book, which totally transformed my attitude towards gardening, agriculture, and farming in general. Until becoming acquainted with David and his work, I had viewed gardening as something my grandfather (and others who couldn't find more interesting hobbies) did—hard work that was totally unnecessary in our modern world. I've always felt like God sends me writing assignments on topics I need to hear, but the paradigm shift I experienced on farming was extreme, to say the least.

David's second book, Counsels on Gardening, is (like this) a compilation of Ellen G. White quotes on the topic of agriculture. The difference is that David's compilation is in chronological order, as he wanted to preserve the historical context of each quote. In contrast, this compilation is arranged by topic, with study questions included. I did draw quite a few quotes from Counsels on Gardening (with David's permission), so some of the quotes will be the same. If you would like to read a more complete compilation and understand the progression of Ellen White's thinking on agriculture set in historical context, I highly recommend Counsels on Gardening. If you desire more of topical reference or syllabus, this compilation (Hope in the Soil) is a great place to start.

In putting together this compilation, I am also deeply indebted to the work of the late **Joe Engelkemier**. Last issued in 1981, Engelkemier's syllabus Training in Soil Cultivation as a Part of Adventist Education served as a syllabus from the Andrews University agricultural program for some years. That syllabus was extremely helpful to me in putting together this compilation.

I would also like to mention the *Michigan Conference of Seventh-day Adventists* and the *Good News Farm* in Cedar Lake, Michigan. From 2014-2015, I had the privilege of helping with the accounting at *Good News Farm*, which opened my eyes to the blessings of organic farm produce, farming as a work opportunity for students (in this case *Great Lakes Adventist Academy*), and the incredible opportunities that farming offers as a training ground for both business and character development.

Until recent years, I had considered myself to be someone with a decidedly non-green thumb and zero interest in gardening or agriculture. I thank God for changing my thinking. It is my sincere desire that this compilation will be a help to others who, like myself, were unaware of the definitive counsel provided by God, through the writings of Ellen G. White, on agriculture.

—*Cari Haus*

Note: *All Bible verses in this compilation are in the King James Version unless otherwise noted.*

The Relation of True Education to the Farm (an article by Ellen G. White)

Published in The Advocate, March 1, 1901

No pains should be spared to select places for our schools where the moral atmosphere will be as healthful as possible, for the influence that prevails will leave a deep impress on young and forming characters. For this reason a retired locality is best. The great cities, the centers of business and learning, may seem to present some advantages, but these advantages are outweighed by other considerations. How many children there are in the crowded cities who have not even a spot of green grass to set their feet upon. If they could be educated in the country, amid the beauty, peace, and purity of nature, it would seem to them the spot nearest heaven. In the retired places, where we are farthest from the corrupting maxims, customs, and excitements of the world, and nearest to the heart of nature, Christ makes his presence real to us, and speaks to our souls of his peace and love. Serious times are before us, and there is great need for the families to get out of the cities into the country. (Christian Schools p. 81)

The youth educated in the large cities are surrounded by influences similar to those that prevailed before the flood. The same principles of disregard for God and his law, the same love of pleasure, of selfish gratification, and of pride and vanity, are at work at the present time. The world is given up to pleasure; immorality prevails; the rights of the weak and helpless are disregarded, and the world over, the large cities are fast becoming hotbeds of iniquity. (Special Testimonies on Education p. 44)

There is room within earth's vast boundaries for schools to be located, where ground can be cleared, land cultivated, and where a proper education can be given. This work is essentially an all-round education, and one which is favorable to spiritual advancement. Nature's voice is the voice of Jesus Christ, teaching us innumerable lessons of perseverance. The mountains and hills are changing, the earth is waxing old like a garment, but the blessing of God which spreads a table for his people in the wilderness, will never pass away. (Christian Schools p. 80)

The children and youth, all classes of students, need the lessons to be derived from this source. In itself, the beauty of nature leads the soul away from sin and worldly attractions, and toward purity, peace, and God. For this reason the cultivation of the soil is good work for children and youth. It brings them into direct contact with nature and nature's God, and that they may have this advantage in connection with our schools, there should be, so far as possible, large flower gardens and extensive lands for cultivation. A return to simpler methods will be appreciated by the children and youth. Work in the garden and field will be an agreeable change from the wearisome routine of abstract lessons to which their young minds should never be confined. (Special Testimonies on Education pp. 60-61)

In the school that is started here in Cooranbong, we look to see real success in agricultural lines, combined with the study of the sciences. We mean this place to be a center from which shall irradiate light, precious advanced knowledge, that shall result in the working of unimproved lands, so that hills and valleys shall blossom as the rose. For both children and men, labor combined with mental taxation will give the right kind of all-round education. The cultivation of the mind will bring tact and fresh incentive to the cultivation of the soil. (Special Testimonies to Ministers and Workers 4:19)

The school has made an excellent beginning. The students are learning how to plant trees, strawberries, etc.; how they must keep every spangle and fiber of the roots uncramped, in order to give them a chance to grow. Is not this a most precious lesson as to how to treat the human mind, and the body as,—not to cramp any of the organs of the body, but to give them ample room to do their work? (Special Testimonies to Ministers and Workers 4:17)

The students are learning what plowing means, and that the hoe and the shovel, the rake and the harrow, are all implements of honorable industry. Mistakes will often be made, but error lies close beside truth. Wisdom will be learned by failures, and the energy that will make a beginning, gives hope of success in the end. Hesitation will keep things back, precipitancy will alike retard, but all will serve as lessons, if the human agents will have it so. (Special Testimonies to Ministers and Workers 4:18)

There will be a new presentation of men as bread-winners, possessing educated, trained ability to work the soil to advantage. Such men will break down the foolish sentiments that have prevailed in regard to manual labor. An influence will go forth, not in loud-voiced oratory, but in real inculcation of ideas. We shall see farmers who are not coarse and rough and slack, careless of their apparel and of the appearance of their homes; but they will bring taste into farm houses. Rooms will be sunny and inviting. We shall not see blackened ceilings, covered with cloth full of dust and dirt. Science, genius, intelligence, will be manifest in the home. The cultivation of the soil will be regarded as elevating and ennobling. (Special Testimonies to Ministers and Workers 4:19)

We should work the soil cheerfully, hopefully, gratefully, believing that the earth holds in her bosom rich stores for the faithful worker to garner, richer than gold or silver. The niggardliness laid to her charge is false witness. With proper, intelligent cultivation, the earth will yield her treasures for the benefit of man. The cultivation of our land requires the exercise of all the brain power and tact we possess. The lands around us testify to the indolence of man. We hope to arouse to action the dormant senses. We hope to see intelligent farmers who will be rewarded for their earnest labor. The hand and heart must co-operate, bringing new and sensible plans into operation in the cultivation of the soil. (Special Testimonies to Ministers and Workers 4:18)

Men take you to their orchards of oranges and lemons and other fruit, and tell you the produce does not pay for the work done on them. It is next to impossible to make ends meet, and parents decide that the children shall not be farmers. They have not the courage and hope to educate them to till the soil. What is needed is schools to educate and train the youth, so that they will know how to overcome this condition of things. There must be education in the sciences, and education in the plans and methods of working the soil. There is hope in the soil, but brain and heart and strength must be brought into the work of tilling it.

There is need of much more extensive knowledge in regard to the preparation of the soil. There is not sufficient breadth of view as to what can be realized from the earth. A narrow and unvarying routine is followed with discouraging results. Let the educated ability be employed in devising improved methods of work. This is just what the Lord wants. There is need of intelligent and educated ability to devise the best methods in farming, in building, and in every other department, that the worker may not labor in vain. God, who has made the world for the benefit of man, will provide means from the earth to sustain the diligent worker. The seed placed in thoroughly prepared soil will produce its harvest. God can spread a table for his people in the wilderness. There is much mourning over unproductive soil, when, if men would read the Old Testament Scriptures, they would see that the Lord knew much better than they in regard to the proper treatment of the land. After being worked for several years, and giving her treasures to the possession of men, portions of the land should be allowed to rest, and then the crops should be changed. We might learn much, also, from the Old Testament, in regard to the labor problem. (Special Testimonies on Education p. 100)

The earth has its concealed treasures, and the Lord would have thousands and tens of thousands working upon the soil, who are crowded into the cities to watch for a chance to earn a trifle. The earth is to be made to give forth its strength, but without the blessing of God it can do nothing. In the beginning, God looked upon all he had made, and pronounced it very good. The curse was brought upon the earth in consequence of sin, but shall this curse be multiplied by increasing sin? Ignorance is doing its baleful work. Slothful servants are increasing the evil by their lazy habits. Many are unwilling to earn their bread by the sweat of their brow, and they refuse to till the soil. But the earth has blessings hidden in her depths for those who have courage and will and perseverance to gather her treasures. (Special Testimonies on Education p. 104)

God would be glorified if men from other countries who have acquired an intelligent knowledge of agriculture, would come to Australia, and by precept and example teach the people how to cultivate the soil, that it may yield rich treasures. Men are wanted to educate others how to plow, and how to use the implements of agriculture. Who will be missionaries to do this work, to teach proper methods to the youth, and to all who feel willing and humble enough to learn? (Special Testimonies on Education p. 101)

—Ellen G. White

Chapter #1: The Key to Unlocking God's Word

Day unto day uttereth speech, and night unto night sheweth knowledge. There is no speech nor language, where their voice is not heard. (Psalm 19:2-3)

KEY TO UNLOCKING GOD'S WORD FOUND IN THE NATURAL WORLD: God has, in the natural world, placed in the hands of the children of men the key to unlock the treasure-house of his word. The unseen is illustrated by the seen; divine wisdom, eternal truth, infinite grace, are understood by the things that God has made. Then let the children and youth become acquainted with nature and nature's laws. Let the mind be developed to the utmost capacity, and the physical powers trained for the practical duties of life; but teach them also that God has made this world fair because he delights in our happiness; and that a more beautiful home is preparing for us in that world where there will be no more sin. The word of God declares, "Eye hath not seen, nor ear heard, neither have entered into the heart of man, the things which God hath prepared for them that love him." (My Life Today p. 112)

THE ENTIRE NATURAL WORLD INTERPRETS THE THINGS OF GOD: The whole natural world is designed to be an interpreter of the things of God. To Adam and Eve in their Eden home, nature was full of the knowledge of God, teeming with divine instruction. To their attentive ears it was vocal with the voice of wisdom. Wisdom spoke to the eye and was received into the heart, for they communed with God in His created works. As soon as the holy pair transgressed the law of the Most High, the brightness from the face of God departed from the face of nature. Nature is now marred and defiled by sin. But God's object lessons are not obliterated; even now, rightly studied and interpreted, she speaks of her Creator.... The children and youth, all classes of students, need the lessons to be derived from this source. In itself the beauty of nature leads the soul away from sin and worldly attractions, and toward purity, peace, and God. (Counsels to Parents, Teachers, and Students pp. 185-186)

NATURE SECOND ONLY TO THE BIBLE AS TESTAMENT OF GOD'S POWER AND LOVE: While the Bible should hold the first place in the education of children and youth, the book of nature is next in importance. God's created works testify to his love and power. He has called the world into being, with all that it contains. God is a lover of the beautiful; and in the world which he has fitted up for us, he has not only given us everything necessary for our comfort, but he has filled the heavens and the earth with beauty. We see his love and care in the rich fields of autumn, and his smile in the glad sunshine. His hand has made the castle-like rocks and the towering mountains. The lofty trees grow at his command; he has spread earth's green velvet carpet, and dotted it with shrubs and flowers. Why has he clothed the earth and trees with living green, instead of a dark, somber brown? Is it not that they may be more pleasing to the eye? And shall not our hearts be filled with gratitude, as we read the evidences of his wisdom and love in the wonders of his creation? (Lift Him Up p. 67)

IN QUIET PLACES GOD BECOMES REAL: God would have us appreciate his blessings in his created works. How many children there are in the crowded cities that have not even a spot of green grass to set their feet upon. If they could be educated in the country, amid the beauty, peace, and purity of nature, it would seem to them the spot nearest heaven. In retired places, where we are farthest from the corrupting maxims, customs, and excitements of the world, and nearest to the heart of nature, Christ makes his presence real to us, and speaks to our souls of his peace and love. (Fundamentals of Christian Education p. 424)

VALUABLE SPIRITUAL LESSONS—AND INSPIRATION—DRAWN FROM NATURE AND VEGETATION: God designs we should draw lessons from nature, and make a practical application of these lessons to our own lives. Although we may suffer under disappointments, reverses, and affliction, yet we cannot afford to fret, and walk under a cloud, and cast a shadow upon all with whom we associate. Invalids may imitate nature. They need not be like a withered, decaying branch. Let vegetation, that is clothed in cheerful green, cheer and comfort you, and suggest to you the happiness that you may reflect upon others, by presenting before them the aspect of freshness and cheerfulness, instead of complaints, sighs, and groans, and apparent languor in every step, and an appearance of inability in every move. (Health Reformer, June 1, 1871)

SOFTENING, SUBDUING INFLUENCE OUT IN THE OPEN AIR: To those who labor in the open air new scenes are continually unfolding. As they make the book of nature their study, a softening, subduing influence comes over them. (Fundamentals of Christian Education p. 319)

NATURE'S BEAUTY BUT A GLEAM SHINING FROM GOD'S GLORY: In the loveliness of the things of nature you may learn more of the wisdom of God than the schoolmen know. On the lily's petals, God has written a message for you, written in language that your heart can read only as it unlearns the lessons of distrust and selfishness and corroding care. Why has He given you the singing birds and the gentle blossoms, but from the overflowing love of a Father's heart, that would brighten and gladden your path of life? All that was needed for existence would have been yours without the flowers and birds, but God was not content to provide what would suffice for mere existence. He has filled earth and air and sky with glimpses of beauty to tell you of His loving thought for you. The beauty of all created things is but a gleam from the shining of His glory. (Mount of Blessing p. 96)

LILIES BLOSSOMING TEACH WHAT GOD CAN DO IN OUR HEARTS: Consider the lilies. Every flower that opens its petals to the sunshine obeys the same great laws that guide the stars, and how simple and beautiful and how sweet its life! Through the flowers, God would call our attention to the loveliness of Christlike character. He who has given such beauty to the blossoms desires far more that the soul should be clothed with the beauty of the character of Christ. Consider, says Jesus, how the lilies grow; how, springing from the cold, dark earth, or from the mud of the river bed, the plants unfold in loveliness and fragrance. Who would dream of the possibilities of beauty in the rough brown bulb of the lily? But when the life of God, hidden therein, unfolds at His call in the rain and the sunshine, men marvel at the vision of grace and loveliness. Even so will the life of God unfold in every human soul that will yield itself to the ministry of His grace, which, free as the rain and the sunshine, comes with its benediction to all. It is the word of God that creates the flowers, and the same word will produce in you the graces of His Spirit. (Mount of Blessing p. 97)

BEAUTIES OF NATURE CARRY THE MIND UP TO THE AUTHOR OF ALL THAT IS LOVELY: The beauties in nature are a theme for contemplation. In studying the natural loveliness surrounding us, the mind is carried up through nature to the Author of all that is lovely. All the works of God are speaking to our senses, magnifying His power, exalting His wisdom. Every created thing has in it charms which interest the child of God and mold his taste to regard these precious evidences of God's love above the work of human skill. The prophet, in words of glowing fervor, magnifies God in His created works: When I consider Thy heavens, the work of Thy fingers, the moon and the stars, which Thou hast ordained; what is man, that Thou art mindful of him? and the son of man, that Thou visitest him? O Lord our Lord, how excellent is Thy name in all the earth! I will praise thee, O Lord, with my whole heart; I will show forth all Thy marvelous works. (Testimonies to the Church Vol. 3 p. 377)

GOD DESIGNED NATURE TO ATTRACT AND INTEREST THE MIND: God has surrounded us with nature's beautiful scenery to attract and interest the mind. It is His design that we should associate the glories of nature with His character. If we faithfully study the book of nature, we shall find it a fruitful source for contemplating the infinite love and power of God. (Adventist Home p. 144)

SPIRITUAL EFFECT OF BEHOLDING GOD IN NATURE: Every child may gain knowledge as Jesus did, from the works of nature and the pages of God's holy word. As we try to become acquainted with our Heavenly Father through His word, angels will come near, our minds will be strengthened, our character will be elevated and refined, and we shall become more like our Savior. And as we behold the beautiful and grand in nature, our affections go out after God; while the spirit is awed, the soul is invigorated by coming in contact with the Infinite through His works. Communion with God through prayer develops the mental and moral faculties, and the spiritual powers strengthen as we cultivate thoughts upon spiritual things. (Fundamentals of Christian Education p. 443)

HEARTS SOFTENED BY THE BEAUTIES OF NATURE: What an influence an outdoor life among the flowers and fruit-laden trees has upon those who are sick both in body and in mind! After they stay for a short time at a sanitarium situated in the midst of the beauties of nature, hope begins to take the place of despair. The heart is softened by the objects of beauty in nature, that the great Master Artist has given to mankind as pictures in which are portrayed His goodness and love... (Medical Ministry p. 232)

Part 1: Nature the Most Effective Way to Reach Minds Far from God

For all flesh is as grass, and all the glory of man as the flower of grass. The grass withereth, and the flower thereof falleth away: but the word of the Lord endureth for ever. (1 Peter 1:24-25)

AN UNDERSTANDING OF GOD'S HAND IN NATURE LEADS TO ACKNOWLEDGMENTS OF HIS POWER: The Lord has given His life to the trees and vines of His creation. His word can increase or decrease the fruit of the land. If men would open their understanding to discern the relation between nature and nature's God, faithful acknowledgments of the Creator's power would be heard. Without the life of God, nature would die. His creative works are dependent on Him. He bestows life-giving properties on all that nature produces. We are to regard the trees laden with fruit as the gift of God, just as much as though He placed the fruit in our hands (Manuscript 114, 1899)

NATURE THE MOST EFFECTIVE TOOL TO REACH THOSE WHO DON'T KNOW GOD: The most effective way to teach the heathen who know not God is through His works. In this way, far more readily than by any other method, they can be made to realize the difference between their idols, the works of their own hands, and the true God, the Maker of heaven and earth.... There is a simplicity and purity in these lessons direct from nature that makes of the highest value to others besides the heathen. The children and youth, all classes of students, need the lessons to be derived from this source. In itself the beauty of nature leads the soul away from sin and worldly attractions, and toward purity, peace, and God. (Counsels to Parents, Teachers and Students p. 186)

FLOWERS AND ORCHARDS PREPARE THE MIND TO APPRECIATE GOD'S WORD: By the influence of the quickening, reviving, life-giving properties of nature's great medicinal resources the functions of the body are strengthened, the intellect awakened, the imagination quickened, the spirits enlivened, and the mind prepared to appreciate the beauty of God's word. (Testimonies to the Church Vol. 7 p. 86)

THINGS OF NATURE ARE GOD'S PREACHERS: If our minds are open to the impressions of the Spirit of God, we may learn lessons from the simple and beautiful things of nature. I feel oppressed in the crowded cities where there is naught for the eyes to look upon but houses. The flowers are to us constant teachers. The shrubs and flowers gather to themselves the properties of earth and air which they appropriate to perfect the beautiful buds and blossoming flowers for our happiness, but they are God's preachers, and we are to consider the lessons which they teach us. (That I May Know Him p. 213)

JESUS COMMUNICATED DIVINE TRUTHS THROUGH ILLUSTRATIONS FROM THE NATURAL WORLD: For His own wise purpose the Lord veils spiritual truths in figures and symbols. Through the use of figures of speech the plainest and most telling rebuke was often given to His accusers and enemies, and they could find in His words no occasion to condemn Him. In parables and comparisons He found the best method of communicating divine truth. In simple language, using figures and illustrations drawn from the natural world, He opened spiritual truth to His hearers, and gave expression to precious principles that would have passed from their minds, and left scarcely a trace, had He not connected His words with stirring scenes of life, experience, or nature. In this way He called forth their interest, aroused inquiry, and when He had fully secured their attention, He decidedly impressed upon them the testimony of truth. In this way He was able to make sufficient impression upon the heart so that afterward His hearers could look upon the thing with which He connected His lesson, and recall the words of the divine Teacher. (Fundamentals of Christian Education p. 236)

THE BEAUTIFES OF NATURE INVITE US TO LEARN ABOUT NATURE'S GOD: The beauties of nature have a tongue that speaks to our senses without ceasing. The open heart can be impressed with the love and glory of God as seen in the works of His hand. The listening ear can hear and understand the communications of God through the works of nature. There is a lesson in the sunbeam and in the various objects in nature that God has presented to our view. The green fields, the lofty trees, the buds and flowers, the passing cloud, the falling rain, the babbling brook, the sun, moon, and stars in the heavens, all invite our attention and meditation, and bid us become acquainted with God, who made them all. The lessons to be learned from the various objects of the natural world are these: They are obedient to the will of their Creator; they never deny God, never refuse obedience to any intimation of His will. Fallen beings alone refuse to yield full obedience to their Maker. (Testimonies to the Church Vol. 3 p. 333)

JESUS USED LESSONS FROM NATURE TO PRESS HOME SPIRITUAL LESSONS: He [Jesus] made use of the lofty trees, the cultivated soil, the barren rocks, the flowers of beauty struggling through the clefts, the everlasting hills, the glowing flowers of the valley, the birds caroling their songs in the leafy branches, the spotless lily resting in purity upon the bosom of the water. All these objects that made up the living scene around them were made the medium by which His lessons were impressed upon the minds of His hearers. They were thus brought home to the hearts of all, ... leading them gently up from the contemplation of the Creator's works in nature to nature's God.... (In Heavenly Places p. 114)

JESUS USED NATURE TO SOFTEN THE HEARTS OF HIS HEARERS: The Great Teacher brought His hearers in contact with nature, that they might listen to the voice which speaks in all created things; and as their hearts became tender and their minds receptive, He helped them to interpret the spiritual teaching of the scenes upon which their eyes rested.... In His lessons there was something to interest every mind, to appeal to every heart. Thus the daily task, instead of being a mere round of toil, bereft of higher thoughts, was brightened and uplifted by constant reminders of the spiritual and the unseen. (Adventist Home p. 144)

THE BOOK OF NATURE, LIKE THE BIBLE, GUIDES LOST SHEEP TO THE FOLD: As divine truth is revealed in Holy Writ, so it is reflected, as from a mirror, in the face of nature; and through his creation we become acquainted with the Creator. And so the book of nature becomes a great lesson book, which instructors who are wise can use, in connection with the Scriptures, to guide lost sheep back to the fold of God. As the works of God are studied, the Holy Spirit flashes conviction into the mind. It is not the conviction which logical reasoning produces; but unless the mind has become too dark to know God, the eye too dim to see him, the ear too dull to hear his voice, a deeper meaning is grasped, and the sublime, spiritual truths of the written word are impressed on the heart. (Special Testimonies on Education p. 59)

THE WAYS OF GOD REVEALED IN THE NATURAL WORLD A TREASURY FROM WHICH WE CAN DRAW: He who has a knowledge of God and His Word has a settled faith in the divinity of the Holy Scriptures. He does not test the Bible by man's ideas of science. He brings these ideas to the test of the unerring standard. He knows that God's Word is truth, and truth can never contradict itself.... The ways of God as revealed in the natural world and in His dealings with man constitute a treasury from which every student in the school of Christ may draw (<u>Testimonies for the Church Vol. 8</u> p. 324-325)

Part 2: A Warning from Sodom

For the invisible things of him from the creation of the world are clearly seen, being understood by the things that are made, even his eternal power and Godhead; so that they are without excuse… (Romans 1:20)

Who knoweth not in all these that the hand of the Lord hath wrought this? In whose hand is the soul of every living thing, and the breath of all mankind. (Job 12:9-10)

For the wisdom of this world is foolishness with God. For it is written, He taketh the wise in their own craftiness. (1 Corinthians 3:19)

A WARNING FROM SODOM: Sodom and Gomorrah were like the Garden of Eden. The Lord had lavished His blessings upon that portion of the earth. Everything was beautiful; everything was lovely; and yet it did not lead people to honor the Giver. Christ Triumphant p. 80)

WORD OF GOD NEEDED TO REACH CORRECT CONCLUSIONS ABOUT SCIENCE AND NATURE: God has permitted a flood of light to be poured upon the world in discoveries in science and art; but when professedly scientific men lecture and write upon these subjects from a merely human standpoint, they will assuredly come to wrong conclusions. The greatest minds, if not guided by the Word of God in their research, become bewildered in their attempts to investigate the relations of science and revelation. The Creator and His works are beyond their comprehension; and because they cannot explain these by natural laws, Bible history is considered unreliable. Those who doubt the reliability of the records of the Old and New Testaments will be led to go a step farther, and doubt the existence of God; and then, having let go their anchor, they are left to beat about upon the rocks of infidelity. Moses wrote under the guidance of the Spirit of God, and a correct theory of geology will never claim discoveries that cannot be reconciled with his statements. The idea that many stumble over, that God did not create matter when He brought the world into existence, limits the power of the Holy One of Israel. (Lift Him Up p. 60)

LIGHT SHINING FROM THE CROSS RIGHTLY ILLUMINES NATURE AND INTERPRETS ITS TEACHINGS: When Adam and Eve in Eden lost the garments of holiness, they lost the light that had illuminated nature. No longer could they read it aright. But for those who receive the light of the life of Christ, nature is again illuminated. In the light shining from the cross, we can rightly interpret nature's teaching. (Lift Him Up p. 255)

ONLY THOSE WHO RECOGNIZE GODS HAND IN NATURE CAN LEARN ITS DEEPEST LESSONS: Those alone who recognize in nature their Father's handiwork, who in the richness and beauty of the earth read the Father's handwriting—those alone learn from the things of nature their deepest lessons, and receive their highest ministry. Only those can fully appreciate the significance of hill and vale, river and sea, who look upon them as an expression of the thought of God, a revelation of the Creator. (To Be Like Jesus p. 244)

MINDS CLOSED TO THE HOLY SPIRIT WILL ALWAYS BE IN DARKNESS IN REGARDS TO SCIENCE AND THE POWER OF GOD: Scientific research will open to the minds of the really wise vast fields of thought and information. They will see God in His works, and will praise Him. He will be to them first and best, and the mind will be centered upon Him. Skeptics, who read the Bible for the sake of caviling, through ignorance claim to find decided contradictions between science and revelation. But man's measurement of God will never be correct. The mind unenlightened by God's Spirit will ever be in darkness in regard to His power. (Lift Him Up p. 61)

THE BIBLE—NOT SCIENCE—THE TRUE TEST OF TRUTH: Many, when they find themselves incapable of measuring the Creator and His works by their own imperfect knowledge of science, doubt the existence of God and attribute infinite power to nature. These persons have lost the simplicity of faith, and are removed far from God in mind and spirit. There should be a settled faith in the divinity of God's holy Word. The Bible is not to be tested by men's ideas of science, but science is to be brought to the test of this unerring standard. When the Bible makes statements of facts in nature, science may be compared with the written Word, and a correct understanding of both will always prove them to be in harmony. One does not contradict the other. All truth, whether in nature or revelation, agrees. (The Signs of the Times, March 13, 1884)

SCIENTISTS WHO DENY GOD'S PERSONHOOD AND CONFINE HIM TO A SET OF FIXED LAWS ARE JUST CHASING BUBBLES: Men of science think that with their enlarged conceptions they can comprehend the wisdom of God, that which He has done or can do. The idea largely prevails that He is bounded and restricted by His own laws. Men either deny and ignore His existence, or think to explain everything, even the operations of His Spirit upon the human heart, by natural laws; and they no longer reverence His name or fear His power. While they think they are gaining everything, they are chasing bubbles, and losing precious opportunities to become acquainted with God. They do not believe in the supernatural, not realizing that the Author of nature's laws can work above those laws. They deny the claims of God, and neglect the interests of their own souls; but His existence, His character, His laws, are facts that the reasoning of men of the highest attainments cannot overthrow.... (Lift Him Up p. 61)

THE ASSUMPTIONS OF WISDOM MADE BY THE WORLD'S SO-CALLED GREAT MEN—OUTSIDE OF CHRIST—TO BE DESPISED: When we have right views of the power, greatness, and majesty of God, and of the weakness of man, we shall despise the assumptions of wisdom made by earth's so-called great men, who have none of Heaven's nobility in their characters. There is nothing for which men should be praised or exalted. There is no reason why the opinions of the learned should be trusted, when they are disposed to measure divine things by their own perverted conceptions. Those who serve God are the only ones whose opinion and example it is safe to follow. A sanctified heart quickens and intensifies the mental powers. A living faith in God imparts energy; it gives calmness and repose of spirit, and strength and nobility of character. (Lift Him Up p. 61)

GOD HAS GIVEN SUFFICIENT EVIDENCE FOR FAITH: Man will be left without excuse. God has given sufficient evidence upon which to base faith if he wish to believe. In the last days the earth will be almost destitute of true faith. Upon the merest pretense, the Word of God will be considered unreliable, while human reasoning will be received, though it be in opposition to plain Scripture facts. Men will endeavor to explain from natural causes the work of creation, which God has never revealed. But human science cannot search out the secrets of the God of heaven, and explain the stupendous works of creation, which were a miracle of Almighty power, any sooner than it can show how God came into existence. (Lift Him Up p. 59)

Part 3: Farming a Consistent Way to Reach Souls through Nature

THE KINGDOM OF HEAVEN MADE PLAIN THROUGH THE NATURAL WORLD OF FARMING: The planting and the sowing, the gathering of the harvest, and the care of the trees, are to be wonderful lessons for all the students. The invisible links which connect the sowing and the reaping are to be studied, and the goodness of God is to be pointed out and appreciated. It is the Lord that gives the virtue and the power to the soil and to the seed. Were it not for His divine agency, combined with human tact and ability, the seed sown would be useless. There is an unseen power constantly at work as man's servant, to feed and to clothe him. The parable of the seed as studied in the daily experience of teacher and student is to reveal that God is at work in nature, and it is to make plain the things of the kingdom of heaven. (Australasian Union Conference Record, *July 31, 1899)*

CULTIVATION OF THE SOIL BRINGS THE YOUNG INTO CLOSE CONNECTION WITH GOD: There is a simplicity and purity in these lessons directly from nature that make them of the highest value to others besides the heathen. The children and youth, all classes of students, need the lessons to be derived from this source. In itself the beauty of nature leads the soul away from sin and worldly attractions, and toward purity, peace, and God. For this reason the cultivation of the soil is good work for children and youth. It brings them into direct contact with nature and nature's God. And that they may have this advantage in connection with our schools there should be, as far as possible, large flower gardens and extensive lands for cultivation. (Special Testimonies on Education p. 60)

DAILY TASKS SHOULD LEAD US CLOSER TO GOD: Christ has linked His teaching, not only with the day of rest, but with the week of toil.... In the plowing and sowing, the tilling and reaping, He teaches us to see an illustration of His work of grace in the heart. So in every line of useful labor and every association of life, He desires us to find a lesson of divine truth. Then our daily toil will no longer absorb our attention and lead us to forget God; it will continually remind us of our Creator and Redeemer. The thought of God will run like a thread of gold through all our homely cares and occupations. For us the glory of His face will again rest upon the face of nature. We shall ever be learning new lessons of heavenly truth and growing into the image of His purity. (Adventist Home p. 144)

STUDENTS SHOULD BE POINTED TO THE SPIRITUAL LESSONS RELATED TO PRACTICAL THINGS: Students in the industrial departments, whether they are employed in domestic work, in cultivating the ground, or in other ways, should have time and opportunity given them to tell the practical, spiritual lessons they have learned in connection with the work. In all the practical duties of life, comparisons should be made with the teachings of nature and of the Bible. (Testimonies to the Church Vol. 6 p. 176)

THE FAITHFUL ONES IN OLD TESTAMENT TIMES LIVED CLOSE TO NATURE, KEEPING FLOCKS AND TILLING THE SOIL: The education centering in the family was that which prevailed in the days of the patriarchs. For the schools thus established, God provided the conditions most favorable for the development of character. The people who were under His direction still pursued the plan of life that He had appointed in the beginning. Those who departed from God built for themselves cities, and, congregating in them, gloried in the splendor, the luxury, and the vice that make the cities of today the world's pride and its curse. But the men who held fast God's principles of life dwelt among the fields and hills. They were tillers of the soil and keepers of flocks and herds, and in this free, independent life, with its opportunities for labor and study and meditation, they learned of God and taught their children of His works and ways. Education p. 33)

GOD TRAINED MOSES AND JOHN THE BAPTIST OUT IN THE WILDERNESS: There is a refining, subduing influence in nature that should be taken into account in selecting the locality for a school. God has regarded this principle in training men for his work. Moses spent forty years in the wilds of Midian. John the Baptist was not fitted for his high calling as the forerunner of Christ by association with the great men of the nation in the schools at Jerusalem. He went out into the wilderness, where the customs and doctrines of men could not mold his mind, and where he could hold unobstructed communion with God. (Special Testimonies on Education 46)

MANY HONORED WORKERS IN GOD'S CAUSE CAME FROM THE HUMBLE PROFESSIONS OF LIFE: How many useful and honored workers in God's cause have received a training amid the humble duties of the most lowly positions in life. Moses was the prospective ruler of Egypt, but God could not take him from the king's court to do the work appointed him. Only when he had been for forty years a faithful shepherd was he sent to be the deliverer of his people. Gideon was taken from the threshing of wheat to be the instrument in the hands of God for delivering the armies of Israel. Elisha was called to leave the plow and do the bidding of God. Amos was husbandman, a tiller of the soil, when God gave him a message to proclaim. (Gospel Workers, 1892 p. 316)

Chapter #2: The Farm as God's Training School

Part 1: Youth in Critical Spiritual Need

IN 1857—MANY OF THE YOUNG UNCONVERTED: August 22, 1857, at the house of prayer in Monterey, Michigan, I was shown that many have not yet heard the voice of Jesus, and the saving message has not taken hold of the soul and worked a reformation in the life. Many of the young have not the spirit of Jesus. The love for God is not in their hearts, therefore all the natural besetments hold the victory instead of the Spirit of God and salvation. (Testimonies to the Church Vol. 1 p. 154)

IN 1867—NOT ONE IN TWENTY ADVENTIST YOUNG PEOPLE CONVERTED: "Young Sabbathkeepers are given to pleasure seeking. I saw that there is not one in twenty who knows what experimental religion is. (Testimonies to the Church Vol. 1 p. 96)

EXPANDED CAUSE FOR ALARM: I saw that there is not one in twenty who knows what experimental religion is. They are constantly grasping after something to satisfy their desire for change, for amusement. Their unconsecrated lives are a reproach to the Christian name; their example is a snare to others. They hinder the sinner, for in nearly every respect they are no better than unbelievers. They have the word of God, but its warnings, admonitions, reproofs, and corrections are unheeded, as are also the encouragements and promises to the obedient and faithful. I feel alarmed as I witness everywhere the frivolity of young men and women who profess to believe the truth. God does not seem to be in their thoughts. Their minds are filled with nonsense. Their conversation is only empty, vain talk. They have a keen ear for music, and Satan knows what organs to excite to animate, engross, and charm the mind so that Christ is not desired. The spiritual longings of the soul for divine knowledge, for a growth in grace, are wanting. (Testimonies to the Church Vol. 1 p. 497)

A SOBERING PORTRAYAL: My soul mourns for the youth who are forming characters in this degenerate age. I tremble for their parents also; for I have been shown that as a general thing they do not understand their obligations to train up their children in the way they should go....But very few of the youth are free from corrupt habits. They are excused from physical exercise to a great degree. Idleness leads to the indulgence of corrupt habits. Industry does not weary and exhaust one-fifth part as much as the pernicious habit of self-abuse. Give your children physical labor, which will call into exercise the nerves and muscles. The weariness attending such labor will lessen their inclination to indulge in vicious habits. Idleness is a curse. It produces licentious habits. Many cases have been presented before me, and as I have had a view of their inner lives, my soul has been sick and disgusted with the rottenheartedness of human beings who profess godliness and talk of translation to heaven. I have frequently asked myself: Whom can I trust? Who is free from iniquity? (Testimonies to the Church Vol. 2 348-349)

FOND OF AMUSEMENT, AVERSE TO WORK: The youth of today are a sure index to the future of society; and as we view them, what can we hope for the future? The majority are fond of amusement and averse to work. (Child Guidance p. 440)

THE SAD STATE OF THE YOUTH: The world is full of young men and women who pride themselves upon their ignorance of any useful labor; and they are, almost invariably, frivolous, vain, fond of display, unhappy, unsatisfied, and too often dissipated and unprincipled. Such characters are a blot upon society and a disgrace to their parents. (Child Guidance p. 347)

FOCUSED ON SEDENTARY CAREERS: It is a popular error with a large class to regard work as degrading; therefore young men are very anxious to educate themselves to become teachers, clerks, merchants, lawyers, and to occupy in almost any position that does not require physical labor. (Child Guidance p. 347)

NOT ONE BOY OR GIRL IN A HUNDRED PURE-MINDED WITH UNTAINTED MORALS: It is a painful fact that there is not one girl in a hundred who is pure-minded, and there is not one boy in a hundred whose morals are untainted. Corruption has taken hold of a large class who pass among men and women as polite gentlemen and beautiful ladies. (Testimonies to the Church Vol. 4 p. 96)

NOTHING THAT DEFILES TO ENTER THE CITY OF GOD: Into the city of God there will enter nothing that defiles. All who are to be dwellers there will here have become pure in heart. In one who is learning of Jesus, there will be manifest a growing distaste for careless manners, unseemly language, and coarse thought. When Christ abides in the heart, there will be purity and refinement of thought and manner. (Mount of Blessing p. 24-25)

Part 2: How the Youth Can Be Reached

Let no man despise thy youth; but be thou an example of the believers, in word, in conversation, in charity, in spirit, in faith, in purity. (1 Timothy 4:12)

PUT STUDENTS WHERE NATURE CAN SPEAK TO THEIR SENSES: Schools should be established where there is as much as possible to be found in nature to delight the senses and give variety to the scenery. Let our students be placed where nature can speak to the senses, and in her voice they may hear the voice of God. Let them be where they can look upon. His wondrous works, and through nature behold her Creator. (Fundamentals of Christian Education p. 320)

CLOSE TO THE LESSON BOOK OF NATURE THE FOUNDATION OF TRUE EDUCATION CAN BE FIRMLY AND SURELY MADE: So far as possible, let the child from his earliest years be placed where this wonderful lesson-book [of nature] shall be open before him. Let him behold the glorious scenes painted by the great Master Artist upon the shifting canvas of the heavens, let him become acquainted with the wonders of earth and sea, let him watch the unfolding mysteries of the changing seasons, and, in all His works, learn of the Creator. In no other way can the foundation of a true education be so firmly and surely laid. Yet even the child, as he comes in contact with nature, will see cause for perplexity. He cannot but recognize the working of antagonistic forces. It is here that nature needs an interpreter. Looking upon the evil manifest even in the natural world, all have the same sorrowful lesson to learn—an enemy hath done this. (Lift Him Up p. 255)

NATURE LEADS STUDENT MINDS TOWARDS PURITY, PEACE, AND GOD: In itself the beauty of nature leads the soul away from sin and worldly attractions, and toward purity, peace, and God. Too often the minds of students are occupied with men's theories and speculations, falsely called science and philosophy. They need to be brought into close contact with nature. Let them learn that creation and Christianity have one

God. Let them be taught to see the harmony of the natural with the spiritual. Let everything which their eyes see or their hands handle be made a lesson in character building. Thus the mental powers will be strengthened, the character developed, the whole life ennobled. (Christ's Object Lessons p. 24)

NATURE CONTINUOUS INFLUENCE AND INVITATION: Every object in nature repeats His invitation, 'Come unto Me, all ye that labor and are heavy laden, and I will give you rest.' Matt. 11:28 (Desire of Ages p. 289)

CULTIVATION OF THE SOIL CONTRIBUTES TO DEVELOPMENT OF PURITY OF HEART: So long as God gives me power to speak to our people, I shall continue to call upon parents to leave the cities and get homes in the country, where they can cultivate the soil and learn from the book of nature the lessons of purity and simplicity. (Adventist Home pp. 146-147)

NATURE AWAKENS THE DESIRE TO BE PURE AND GUILELESS, AND PREPARES THE MIND TO APPRECIATE GOD'S WORD: From out-of-door life, men, women and children will gain the desire to be pure and guileless. By the influence of the quickening, reviving, life-giving properties of nature's great medicinal resources, the functions of the body are strengthened, the intellect awakened, the imagination quickened, the spirits enlivened. The mind is prepared to appreciate the beauties of God's Word. (Sons and Daughters of God p. 170)

PRE-OCCUPYING THE MIND WITH GOOD A MOST POWERFUL SAFEGUARD: As safeguard against evil, the preoccupation of the mind with good is worth more than unnumbered barriers of law and discipline. Education p. 213)

NEARLY 100% CONVERSION RATE AT THE AVONDALE SCHOOL: Every term of school which we have held at Avondale has resulted in the conversion of nearly every student in the school. In some terms this has been the case without exception, and in others there have not been more than two or three exceptions. (General Conference Bulletin, April 14, 1901) (*Note: The Avondale School, the Australian college mentioned in this quote, was patterned after the model school plan—with a large and successful farm as an integral part of the educational program*).

GOD CAN DO MUCH FOR THE YOUTH: What shall be done to save our youth? We can do little, but God lives and reigns, and He can do much. The youth are our hope for missionary labor. (Fundamentals of Christian Education p. 320)

Part 3: Education Close to Nature Superior for Spiritual Growth

EDUCATION CLOSE TO NATURE BENEFICIAL FOR SPIRITUAL ADVANCEMENT: Look at nature. There is room in her vast boundaries for schools to be located where grounds can be cleared, land cultivated, and where a proper education can be given. This work is essential for an all-round education, and one which is favorable to spiritual advancement. Nature's voice is the voice of Jesus Christ teaching us innumerable lessons of perseverance. The mountains and hills are changing, the earth is waxing old like a garment, but the blessing of God, which spreads a table for His people in the wilderness, will never cease. (Manuscript Release 8 p. 200)

SCHOOLS CLOSE TO NATURE THE MOST FAVORABLE FOR SPIRITUAL ADVANCEMENT: It reveals cowardice to move so slowly and uncertainly in the labor line—that line which will give the very best kind of education. Look at nature. There is room within her vast boundaries for schools to be established where grounds can be cleared and land cultivated. This work is essential to the education most favorable to spiritual advancement; for nature's voice is the voice of Christ, teaching us innumerable lessons of love and power and submission and perseverance. Some do not appreciate the value of agricultural work. (Testimonies to the Church Vol. 6 178)

FREEDOM FROM MANY TEMPTATIONS FOUND IN THE COUNTRY: Our retired location will offer comparative freedom from many of the temptations of city life. Here are no liquor-selling hotels or dram-shops on every corner to tempt the unfortunate victim of intemperance. And the pure sights and sounds, the clear, invigorating air, and the sense of God's presence pervading all nature, tend to uplift the mind, to soften the heart, and to strengthen the will to resist temptation. (Manuscript Release 11 p. 221)

THE THINGS OF NATURE ARE GOD'S SILENT MINISTERS, MEANT TO TEACH US SPIRITUAL TRUTHS: So long as God gives me power to speak to our people, I shall continue to call upon parents to leave the cities and get homes in the country, where they can cultivate the soil and learn from the book of nature the lessons of purity and simplicity. The things of nature are the Lord's silent ministers, given to us to teach us spiritual truths. They speak to us of the love of God and declare the wisdom of the great Master Artist. (Adventist Home p. 146)

BELIEVERS TO MOVE TO THE COUNTRY TO SAVE THEIR CHILDREN FROM RUIN: Believers who are now living in the cities will have to move to the country, that they may save their children from ruin. Attention must be given to the establishment of industries in which these families can find employment. Those who have charge of the schoolwork at——and——should see what can be done by these institutions to establish such industries, so that our people desiring to leave the cities, can obtain modest homes without a large outlay of means, and can also find employment. In both——and——there are favorable and encouraging features for the development of this plan. Study what these features are. (Country Living p. 19)

Part 3: Nature Especially Valuable in Teaching Small Children

NATURE AN UNFAILING SOURCE OF DELIGHT AND INSTRUCTION TO CHILDREN: To the little child, not yet capable of learning from the printed page or of being introduced to the routine of the schoolroom, nature presents an unfailing source of instruction and delight. The heart not yet hardened by contact with evil is quick to recognize the Presence that pervades all created things. The ear as yet undulled by the world's clamor is attentive to the Voice that speaks through nature's utterances. And for those of older years, needing continually its silent reminders of the spiritual and eternal, nature's teaching will be no less a source of pleasure and of instruction. (Child Guidance p. 45)

LITTLE CHILDREN SHOULD BE ESPECIALLY CLOSE TO NATURE: The little children should come especially close to nature. Instead of putting fashion's shackles upon them, let them be free like the lambs, to play in the sweet, fresh sunlight. Point them to shrubs and flowers, the lowly grass and the lofty trees, and let them become familiar with their beautiful, varied, and delicate forms. Teach them to see the wisdom and love of

God in his created works; and as their hearts swell with joy and grateful love, let them join the birds in their songs of praise. (Special Testimonies on Education p. 62)

SPIRITUAL LESSONS FOR CHILDREN FROM FLOWERS AND OTHER PLANTS: While the children and youth gain a knowledge of facts from teachers and textbooks, let them learn to draw lessons and discern truth for themselves. In their gardening, question them as to what they learn from the care of their plants. As they look on a beautiful landscape, ask them why God clothed the fields and woods with such lovely and varied hues. Why was not all colored a somber brown? When they gather the flowers, lead them to think why He spared us the beauty of these wanderers from Eden. Teach them to notice the evidences everywhere manifest in nature of God's thought for us, the wonderful adaptation of all things to our need and happiness. (To Be Like Jesus p. 244)

FATHERS TO IMPRESS THEIR CHILDREN WITH LESSONS FROM FLOWERS: Upon returning home from his business he should find it a pleasant change to spend some time with his children. He may take them into the garden, and show them the opening buds, and the varied tints of the blooming flowers. Through such mediums he may give them the most important lessons concerning the Creator, by opening before them the great book of nature, where the love of God is expressed in every tree, and flower, and blade of grass. He may impress upon their minds the fact that if God cares so much for the trees and flowers, He will care much more for the creatures formed in His image. He may lead them early to understand that God wants children to be lovely, not with artificial adornment, but with beauty of character, the charms of kindness and affection, which will make their hearts bound with joy and happiness. (Reflecting Christ p. 174)

CHILDREN CAN BE TAUGHT VALUABLE SPIRITUAL LESSONS FROM NATURE: Consider the lilies of the field, how they grow; they toil not, neither do they spin; and yet I say unto you, that even Solomon, in all his glory, was not arrayed like one of these. Let the mother take her children with her into the field or garden, and from the things of nature draw lessons that will point them to nature's God, and aid them in the struggle against evil. Let her point them to the lofty trees, the shrubs, and the carpet of green that covers the earth. Let her teach them how the lily, striking its roots down deep through the mire into the sand below, gains nourishment that enables it to send up a pure, beautiful blossom. Then let her show them how, by rejecting that which is impure, and choosing that which is pure, they may grow up into pure, noble men and women. . . . The children need to be given lessons that will nurture in them courage to resist evil. Point them from nature to nature's God, and they will thus become acquainted with the Creator. (Manuscript Release 3 p. 114)

OPEN THE GREAT BOOK OF NATURE TO CHILDREN: Parents, what kind of education are you giving your children? Are you teaching them to cherish that which is pure and lovely, or are you seeking to place their hands in that of the world? Are you spending time and means that they may learn the outward proprieties of life, and secure the superficial, the deceptive adornments of the world? From their earliest childhood, open before them the great book of nature. Teach them the ministry of the flowers. Show them that if Jesus had not come to earth and died, we should have had none of the beautiful things which we now enjoy. Call their attention to the fact that the color and even the arrangement of every delicate bud and flower is an expression of the love of God to human beings, and that affection and gratitude to their heavenly Father should be awakened in their hearts for all these gifts. (To Be Like Jesus p. 235)

THE FIRST SCHOOL FOR CHILDREN SHOULD BE LARGELY IN NATURE: Children should not be long confined within doors, nor should they be required to apply themselves closely to study until a good foundation has been laid for physical development. For the first eight or ten years of a child's life the field or garden is the best schoolroom, the mother the best teacher, nature the best lesson book. Even when the child is old enough to attend school, his health should be regarded as of greater importance than a knowledge of books. He should be surrounded with the conditions most favorable to both physical and mental growth. Education p. 208)

A LIFE CLOSE TO NATURE MUCH MORE FAVORABLE FOR SPIRITUAL DEVELOPMENT OF CHILDREN: Those who will take their families into the country, place them where they have fewer temptations. The children who are with parents that love and fear God, are in every way much better situated to learn of the Great Teacher, who is the source and fountain of wisdom. They have a much more favorable opportunity to gain a fitness for the kingdom of heaven. Send the children to schools located in the city, where every phase of temptation is waiting to attract and demoralize them, and the work of character building is tenfold harder for both parents and children. (Special Testimonies on Education p. 103)

OUTDOOR WORK AN AGREEABLE CHANGE FROM WEARISOME LESSONS: A return to simple methods will be appreciated by the children and youth. Work in the garden and field will be an agreeable change from the wearisome routine of abstract lessons. (Pamphlet 124 p. 15)

CHILDREN SHOULD STUDY NATURE AND THE BIBLE TOGETHER: Children should be encouraged to search out in nature the objects that illustrate Bible teachings, and to trace in the Bible the similitudes drawn from nature. They should search out, both in nature and in Holy Writ, every object representing Christ, and those also that He employed in illustrating truth. Thus may they learn to see Him in tree and vine, in lily and rose, in sun and star. They may learn to hear His voice in the song of birds, in the sighing of the trees, in the rolling thunder, and in the music of the sea. And every object in nature will repeat to them His precious lessons. (Child Guidance p. 46)

Part 4: Farming Invaluable in Training Evangelists

LESSONS LEARNED IN CULTIVATING LAND TO BE APPLIED TO CULTIVATING THE HEART: Wise plans are to be laid for the cultivation of the land. The students are to be given a practical education in agriculture. This education will be of inestimable value to them in their future work. Thorough work is to be done in cultivating the land, and from this the students are to learn how necessary it is to do thorough work in cultivating the garden of the heart. (Manuscript Release 2 p. 68)

LESSONS LEARNED IN TILLING HARD SOIL APPLY TO THE HARDENED HEARTS THAT MUST BE BROKEN UP SO THE WORD OF GOD CAN BE SOWN: Ye are God's husbandry. Will the students apply this lesson while they are working upon the land, tilling the soil, plowing and harrowing, putting all the skill they possess into the work of bringing the land into a condition where it will be fit for the planting of the seed, and the trees, preparatory for the harvest? Will they bear in mind that they are God's husbandry, a part of the Lord's farm, and that in this term of school there is a great deal of work to be done by those who are appointed to watch for souls as they that must give an account? There are hearts that need much more labor bestowed upon them because the soil has not been under the plow or the harrow. The hardened soil must be broken up and subdued, so that the Word of God, the gospel seed, may find favorable soil for the production of a harvest. (Manuscript Release 11 p. 37)

BREAKING THE FALLOW GROUND OF HUMAN HEARTS: Sow for yourselves righteousness; reap in mercy; break up your fallow ground, for it is time to seek the Lord, till He comes and rains righteousness on you. Hosea 10:12, NKJV I want to exhort those who are in positions of responsibility to waken to their duty, and not imperil the cause of present truth by engaging inefficient men and women to do the work of God. We want those who are willing to go into new fields, and to do hard service for the Lord. I remember visiting in Iowa when the country was new, and I saw the farmers breaking the new ground. I noticed that they had heavy teams, and made tremendous efforts to make deep furrows, but the laborers gained strength and muscle by the exercise of their physical powers. It will make our young workers strong to go into new fields, and break up the fallow ground of human hearts. This work will drive them nearer to God. It will help them to see that they are altogether inefficient in themselves. (To Be Like Jesus p. 255)

LESSONS FROM THE WORK OF FARMERS APPLIES TO THE WORK OF CULTIVATING HEARTS: The work of cultivating the heart is profitable at all times and in all places. Ye are God's husbandry, ye are God's building (1 Corinthians 3:9). We may learn a lesson from the work of the farmer in cultivating the field. He must cooperate with God. His part is to prepare the ground, and plant the seed, at the right time and in the right way. God gives the seed life. He sends the sunshine and the showers, and the seed springs up, first the blade, then the ear, after that the full corn in the ear (Mark 4:28). If the farmer fails to do his part, if the human agent does not cooperate with divine agencies, the sun may shine, the dew and the showers may fall upon the soil, but there will be no harvest. And though the work of planting had been done, unless God sent the sunshine and the dew and the rain, the seed would never, never spring up and grow. (Manuscript Release 1 p. 345)

LESSONS ABOUT SPIRITUAL SEEDS PLANTED AND GROWTH LEARNED IN TILLING THE GROUND: The spiritual lessons to be learned are of no mean order.

The seeds of truth sown in the soil of the heart will not all be lost, but will spring up, first the blade, then the ear, then the full corn in the ear. God said in the beginning, Let the earth bring forth grass, the herb yielding seed, and the fruit tree yielding fruit. God created the seed as He did the earth, by the divine word. We are to exercise our reasoning power in the cultivation of the earth, and to have faith in the word of God that has created the fruit of the earth for the service of man. (Manuscript Release 11 p. 184)

THE SUCCESS WE CAN EXPERIENCE IN RESTORING DISEASED LAND TO HEALTH AND REAPING A HARVEST RESEMBLES WHAT GOD CAN DO IN OUR HEARTS: Men were to cooperate with God in restoring the diseased land to health, that it might be a praise and a glory to His name. And as the land they possessed would, if managed with skill and earnestness, produce its treasures, so their hearts, if controlled by God, would reflect His character.... (Bible Commentary Vol. 1 p. 1112)

IN THE WORK OF SAVINGS SOULS WE MUST SOW BESIDE ALL WATERS: We are to sow beside all waters, keeping our souls in the love of God, working while it is day, using the means entrusted to us in the Master's service. Whatever our hands find to do, we are to do it with cheerfulness; whatever sacrifice we are called upon to make, we are to make it cheerfully. As we sow beside all waters, we shall realize the truth of the words, He which soweth bountifully shall reap also bountifully. (The Review and Herald, January 5, 1905)

SPIRITUAL LESSONS FROM THE SOWER WHO WENT FORTH TO SOW: Behold, a sower went forth to sow. (Matthew 13:3). In the East the state of affairs was so unsettled, and there was so great danger from violence that the people dwelt chiefly in walled towns, and the husbandmen went forth daily to their labor outside the walls. So Christ, the heavenly Sower, went forth to sow. He left His home of security and peace, left the glory that He had with the Father before the world was, left His position upon the throne of the universe. He went forth, a suffering, tempted man; went forth in solitude, to sow in tears, to water with His blood, the seed of life for a world lost. His servants in like manner must go forth to sow. When called to become a sower of the seed of truth, Abraham was bidden, Get thee out of thy country, and from thy kindred, and from thy father's house, unto a land that I will shew thee (Genesis 12:1). And he went out, not knowing whither he went (Hebrews 11:8).... In toil and tears, in solitude, and through sacrifice, must the seed be sown. The sower soweth the word. Christ came to sow the world with truth. Ever since the fall of man, Satan has been sowing the seeds of error. It was by a lie that he first gained control over men, and thus he still works to overthrow God's kingdom in the earth and to bring men under his power. A sower from a higher world, Christ came to sow the seeds of truth. He who had stood in the councils of God, who had dwelt in the innermost sanctuary of the Eternal, could bring to men the pure principles of truth. Ever since the fall of man, Christ had been the Revealer of truth to the world. By Him the incorruptible seed, the word of God, which liveth and abideth for ever, is communicated to men (1 Peter 1:23). In that first promise spoken to our fallen race in Eden, Christ was sowing the gospel seed. But it is to His personal ministry among men, and to the work which He thus established, that the parable of the sower especially applies. The word of God is the seed. Every seed has in itself a germinating principle. In it the life of the plant is enfolded. So there is life in God's word. Christ says, The words that I speak unto you, they are spirit, and they are life (John 6:63).... In every command and in every promise of the word of God is the power, the very life of God, by which the command may be fulfilled and the promise realized. He who by faith receives the word is receiving the very life and character of God. Every seed brings forth fruit after its kind. Sow the seed under right conditions, and it will develop its own life in the plant. Receive into the soul by faith the incorruptible seed of the word, and it will bring forth a character and a life after the similitude of the character and the life of God (Christ's Object Lessons pp. 36-38)

KNOWLEDGE OF USEFUL INDUSTRIES HELPFUL IN BREAKING DOWN PREJUDICE: In establishing our schools out of the cities, we shall give the students an opportunity to train the muscles to work as well as the brain to think. Students should be taught how to plant, how to gather the harvest, how to build, how to become acceptable missionary workers in practical lines. By their knowledge of useful industries they will often be enabled to break down prejudice; often they will be able to make themselves so useful that the truth will be recommended by the knowledge they possess. (Counsels to Parents, Teachers and Students p. 309)

MISSIONARIES TRAINED ALONG PRACTICAL LINES WILL BE MOST EFFECTIVE: The usefulness learned on the school farm is the very education that is most essential for those who go out as missionaries to many foreign fields. If this training is given with the glory of God in view, great results will be seen. No work will be more effectual than that done by those who, having obtained an education in practical life, go forth to mission fields with the message of truth, prepared to instruct as they have been instructed. The knowledge they have obtained in the tilling of the soil and other lines of manual work, and which they carry with them to their fields of labor, will make them a blessing even in heathen lands. (Counsels to Parents, Teachers and Students p. 534)

MISSIONARIES WHO KNOW WHO TO COOK AND SEW AND BUILD WILL BE MUCH MORE SUCCESSFUL. AS INDUSTRIAL EDUCATORS, THEY WILL ALSO NEED LESS SUPPORT: Culture on all these points will make our youth useful in carrying the truth to foreign countries. They will not then have to depend upon the people among whom they are living to cook and sew and build for them, nor will it be necessary to spend money to transport men thousands of miles to plan schoolhouses, meetinghouses, and cottages. Missionaries will be much more influential among the people if they are able to teach the inexperienced how to labor according to the best methods and to produce the best results. They will thus be able to demonstrate that missionaries can become industrial educators, and this kind of instruction will be appreciated especially where means are limited. A much smaller fund will be required to sustain such missionaries, because, combined with their studies, they have put to the very best use their physical powers in practical labor; and wherever they may go all they have gained in this line will give them vantage ground. (Testimonies to the Church Vol. 6 p. 176) (*Note: while this quote doesn't apply specifically to farming, the principle is the same.*)

THE EARNESTNESS, DILIGENCE AND PERSEVERING LABOR NEEDED TO TILL THE SOIL MUST BE APPLIED TO THE SOIL OF HEARTS: In tilling the soil, in disciplining and subduing the land, lessons may constantly be learned. No one would think of settling upon a raw piece of land, expecting it at once to yield a harvest. Earnestness, diligence, and persevering labor are to be put forth in treating the soil preparatory to sowing the seed. So it is in the spiritual work in the human heart. Those who would be benefited by the tilling of the soil must go forth with the word of God in their hearts. They will then find the fallow ground of the heart broken by the softening, subduing influence of the Holy Spirit. Unless hard work is bestowed on the soil, it will not yield a harvest. So with the soil of the heart: the Spirit of God must work upon it to refine and discipline it before it can bring forth fruit to the glory of God. (Christ's Object Lessons p. 88)

MAN MAY WORK THE SOIL, BUT IT IS GOD THAT GIVES THE INCREASE. SO IT IS WITH THE HUMAN HEART. Holiness, which means wholeness to God, is wholly acceptable to God. A Paul may plant, and Apollos water, but God giveth the increase. He that soweth to his flesh, shall of the flesh reap corruption; but he that soweth to the Spirit, shall of the Spirit reap life everlasting [Galatians 6:8]. As the workers till the soil, they are to reap all the advantages possible by making an application of the lessons they receive. In the natural world unseen agencies are constantly at work to produce the essential results, but the harvest to be reaped depends upon the seed that has been sown. After man faithfully prepares the land, and plants the seed, God must work constantly to cause the seed to germinate. (Manuscript Release 18 p. 6)

Part 5: God's Ownership of School Farms

LAND AROUND THE SCHOOL IS THE LORD'S FARM: On several occasions the light has come to me that the land around our school is to be used as the Lord's farm. In a special sense portions of this farm should be highly cultivated. Spread out before me I saw land planted with every kind of fruit tree that will bear fruit in this locality; there were also vegetable gardens, where seeds were sown and cultivated. (Testimonies to the Church Vol. 6 p. 185)

GOD'S OWN HAND SEEN IN THE SETTING AROUND ONE SCHOOL: Here is God's great farm. My mind is filled with awe as I look at these giant trees, and consider the fact that this is God's great forest garden which His own hand has planted and cared for, in promoting the growth of trees, shrubs, and beautiful ferns. God's own work is seen in the streams of water on either side of the land purchased for the school. (Manuscript Release 13 p. 353)

THE SCHOOL IS THE LORD'S PROPERTY, AND THE GROUNDS AROUND IT ARE HIS FARM: Then let everything not essential to the work of the school be kept at a distance, and thus prevent any disturbance of the sacredness of the place through the proximity of families and buildings. Let the school stand alone. There must not be this one and that one claiming personal property near it. It will be better for private families, however devoted they may be in the service of the Lord, to be located at some distance from the school buildings. The school is the Lord's property, and the grounds about it are His farm, where the great Sower can make His garden a lesson book. The results of the labors will be seen, first the blade, then the ear, then the full corn in the ear. The land will yield its treasures, bringing the joyousness of an abundant harvest, and the produce gathered through the blessing of God is to be used as nature's lesson book from which spiritual lessons can be made plain, and applied to the necessities of the soul. (Australasian Union Conference Record, July 31, 1899)

GOD'S FARM IS SACRED GROUND: The open book of nature is to be the student's study. Schools should be established away from the cities. I have more invested in this land than any other person. I am carrying students through school, paying their expenses that they may get a start. This gives me an influence with teachers and learners. The land was laid out in lots. Houses were to be built, as in a village. But I tell them that buildings are not to be crowded upon the land around the school buildings. This is God's farm, and it is sacred ground. Here the students are to learn the lesson, Ye are God's husbandry; ye are God's building. The work that is done in the land is to be done in a particular, thorough, wise manner. From the cultivation of the soil and the planting of seed, lessons in spiritual lines may be learned. (Manuscript Release p. 186)

THE FARM—AS WELL AS THE SCHOOL—IS A PRECIOUS TREASURE FROM GOD: As believers in the greatest truths ever given to mortals, we should put to the highest use the talents that God has entrusted to us. The farm and the school at Huntsville have been placed in our hands as a precious treasure. We cannot express in words all that is involved in the proper cultivation of the land and the education of the students in domestic duties. If this work is done in the fear of God, souls will be influenced to take their position on the side of an unpopular truth. (A Place Called Oakwood p. 120)

EDUCATION ON THE SCHOOL FARM PREPARES FOR THE LORD'S FARM IN THE EARTH MADE NEW: Let the students be out in the most healthful location that can be secured, to do the very work that should have been done years ago. Then there would not be so great discouragements. Had this been done, you would have had some grumbling from students, and many objections would have been raised by parents, but this all-round education would educate the children and youth, not only for practical work in various trades, but would prepare them for the Lord's farm in the earth made new. (Manuscript Release 8 p. 199)

THE LAND IS A PRECIOUS TREASURE, A LESSON BOOK TO THE STUDENTS: I am deeply interested in the work that is being done in the Southern field, and especially in the work of the Huntsville School. This school farm was represented to me as having on it fruit trees in full bearing, and also a variety of grains and vegetables, which were in a flourishing condition. Then the words were spoken: This land is a precious treasure. If thoroughly cultivated, it will yield a valuable increase for the support of the school. But special pains must be taken in its cultivation. Much more may be realized from it than now appears possible. If properly treated, this land will be a lesson book to the students, and to our people, and to those not of our faith. (A Place Called Oakwood p. 104)

AVONDALE TO BE THE LORD'S FARM: On several occasions the light has come to me that Avondale is to be used as the Lord's farm. In a special sense there is to be connected with this farm land that shall be highly cultivated. Spread out before me there was land planted with every kind of fruit trees that will bear fruit in this locality, also vegetable gardens, where seeds were sown and cultivated. (Australasian Union Conference Record, July 31, 1899)

THE SCHOOL FARM IS GOD'S LESSON BOOK: The educational advantages of our school are to be of a distinct order. This school farm is God's lesson book. Those who till the soil and plant and cultivate the orchard are to make the application of nature's lessons, and bring these lessons learned into their actual spiritual experience. Let every individual bear in mind that whatsoever a man soweth, that shall he also reap. The man who day by day sows objectionable seeds, in words, in deportment, in spirit, is conforming himself to the same character, and this is determining the future harvest he will reap. (Manuscript Release 18 p. 2)

Chapter #3: Godly Farms—Schools and Family-owned—as Beacons of Light

Part 1: God's Original Homesteading Plan

And the Lord God took the man, and put him into the Garden of Eden to dress it and to keep it. (Genesis 2:15)

— (1): A Plan Developed Before the World Was

A PLAN DEVELOPED BEFORE THE WORLD WAS: When in counsel with the Father before the world was, it was designed that the Lord God should plant a garden for Adam and Eve in Eden and give them the task of caring for the fruit trees and cultivating and training the vegetation. Useful labor was to be their safeguard, and it was to be perpetuated through all generations to the close of earth's history. (Child Guidance p. 345)

AT CREATION, WISDOM SPOKE TO THE EYE AND CREATION EXPRESSED THE VERY THOUGHTS OF GOD: In their original perfection all created things were an expression of the thought of God. To Adam and Eve in their Eden home nature was full of the knowledge of God, teeming with divine instruction. Wisdom spoke to the eye and was received into the heart; for they communed with God in His created works.... The earth is now marred and defiled by sin. Yet even in its blighted state much that is beautiful remains. (The Faith I Live By p. 25)

ALL NATURE ILLUMINATED—THE WORLD A LESSON BOOK—LIFE A SCHOOL: To him who learns thus to interpret its teachings, all nature becomes illuminated; the world is a lesson book, life a school. The unity of man with nature and with God, the universal dominion of law, the results of transgression, cannot fail of impressing the mind and molding the character. These are lessons that our children need to learn. (Child Guidance p. 55)

GOD'S ORIGINAL PLAN TO EDUCATE ADAM: When Adam came from the Creator's hand, he bore, in his physical, mental, and spiritual nature, a likeness to his Maker. God created man in his own image (Genesis 1:27), and it was His purpose that the longer man lived the more fully he should reveal this image—the more fully reflect the glory of the Creator. All his faculties were capable of development; their capacity and vigor were continually to increase. Vast was the scope offered for their exercise, glorious the field opened to their research. The mysteries of the visible universe—the wondrous works of him which is perfect in knowledge (Job 37:16)—invited man's study. Face-to-face, heart-to-heart communion with his Maker was his high privilege. Had he remained loyal to God, all this would have been his forever. Throughout eternal ages he would have continued to gain new treasures of knowledge, to discover fresh springs of happiness, and to obtain clearer and yet clearer conceptions of the wisdom, the power, and the love of God. More and more fully would he have fulfilled the object of his creation, more and more fully have reflected the Creator's glory. (God's Amazing Grace p. 343)

EARTH'S ENTIRE HISTORY LIES BETWEEN TWO EDEN SCHOOLS: Between the school established in Eden at the beginning and the school of the hereafter there lies the whole compass of this world's history—the history of human transgression and suffering, of divine sacrifice, and of victory over death and sin. Not all the conditions of the first school of Eden will be found in the school of the future life. No tree of knowledge of good and evil will afford opportunity for temptation. No tempter is there, no possibility of wrong. Every character has withstood the testing of evil, and none are longer susceptible to its power.... (God's Amazing Grace p. 365)

THE GARDEN OF EDEN—PAST AND FUTURE—WERE AND WILL BE BRANCHES OF THE HEAVENLY SCHOOL: Heaven is a school; its field of study, the universe; its teacher, the Infinite One. A branch of this school was established in Eden; and the plan of redemption accomplished, education will again be taken up in the Eden school.... (God's Amazing Grace p. 365)

THE EDEN HOME: God gave Adam and Even employment. Eden was the school for our first parents, and God was their instructor. They learned how to till the soil and to care for the things which the Lord had planted. (Fundamentals of Christian Education p. 314)

ALMOST LIKE HEAVEN: If the poor now crowded into the cities could find homes upon the land, they might not only earn a livelihood, but find health and happiness now unknown to them. Hard work, simple fare, close economy, often hardship and privation, would be their lot. But what a blessing would be theirs in leaving the city, with its enticements to evil, its turmoil and crime, misery and foulness, for the country's quiet and peace and purity. To many of those living in the cities who have not a spot of green grass to set their feet upon, who year after year have looked out upon filthy courts and narrow alleys, brick walls and pavements, and skies clouded with dust and smoke,—if these could be taken to some farming district, surrounded with the green fields, the woods and hills and brooks, the clear skies and the fresh, pure air of the country, it would seem almost like heaven. Cut off to a great degree from contact with and dependence upon men, and separated from the world's corrupting maxims and customs and excitements, they would come nearer to the heart of nature. God's presence would be more real to them. Many would learn the lesson of dependence upon Him. Through nature they would hear His voice speaking to their hearts of His peace and love, and mind and soul and body would respond to the healing, life-giving power. (Ministry of Healing pp. 190-192)

LAND TO SERVE AS AN OBJECT LESSON: The light given me from the Lord is that whatever land we occupy is to have the very best kind of care and to serve as an object lesson to the colonials of what the land will do if properly worked. (Manuscript Releases Vol. 8 p. 147)

GOD'S PLAN FOR FAMILY LIFE IN ISRAEL: The land of Canaan was divided among the whole people, the Levites only, as ministers of the sanctuary, being exempted…by the distribution of the land among the people, God provided for them, as for the dwellers in Eden, the occupation most favorable to development—the care of plants and animals. Education p. 43)

GOD PROVIDED ISRAEL WITH THE OCCUPATIONS MOST FAVORABLE FOR DEVELOPMENT—THE CARE OF PLANTS AND ANIMALS: By the distribution of the land among the people, God provided for them, as for the dwellers in Eden, the occupation most favorable to development—the care of plants and animals. A further provision for education was the suspension of agricultural labor every seventh year, the land lying fallow, and its spontaneous products being left to the poor. Thus was given opportunity for more extended study, for social interaction and worship, and for the exercise of benevolence, so often crowded out by life's cares and labors. (To Be Like Jesus p. 250)

A PLAN EMPHASIZED AGAIN AND AGAIN: Again and again the Lord has instructed that our people are to take their families away from the cities, into the country, where they can raise their own provisions. (Selected Messages Vol. 2 p. 141)

THE FAMILY HERE TO BE A MODEL OF THE FAMILY IN HEAVEN, THE HOME TO EXEMPLIFY OUR BELIEFS: Parents are under obligation to God to make their surroundings such as will correspond to the truth they profess. They can then give correct lessons to their children, and the children will learn to associate the home below with the home above. The family here must, as far as possible, be a model of the one in heaven. Then temptations to indulge in what is low and groveling will lose much of their force. Children should be taught that they are only probationers here, and educated to become inhabitants of the mansions which Christ is preparing for those who love Him and keep His commandments. This is the highest duty which parents have to perform. (Adventist Home p. 146)

A PLOT OF GROUND FOR EVERY FAMILY: Every family should have a plot of ground for cultivation and beauty. (The Health Reformer, March, 1871)

ALL SHOULD DO MORE TO CULTIVATE THE LAND: Families and institutions should learn to do more in the cultivation of land. All should be acquainted with the special value of fruits and vegetables fresh from the orchard and garden. (Child Guidance p. 357)

A MESSAGE TO FATHERS: You should help your children to , and then assist them in planting their seeds and shrubs. Fathers should take an interest in these things for the benefit of their children, if they themselves have not a natural love for them. (The Health Reformer, March, 1871)

PROVISIONS FOR PROPERTY OWNERSHIP IN THE HEBREW ECONOMY: In apportioning the inheritance of His people, it was God's purpose to teach them, and through them the people of after generations, correct principles concerning the ownership of the land. The land of Canaan was divided among the whole people, the Levites only, as ministers of the sanctuary, being excepted. Though one might for a season dispose of his possession, he could not barter away the inheritance of his children. When able to do so, he was at liberty at any time to redeem it; debts were remitted every seventh year, and in the fiftieth, or year of jubilee, all landed property reverted to the original owner. Thus every family was secured in its possession, and a safeguard was afforded against the extremes either of wealth or of poverty. (Education p. 43)

THE CENTRAL THEME OF THE BIBLE IS THE RESTORATION OF MAN. WE CAN LIVE IN THE ATMOSPHERE OF HEAVEN NOW. The central theme of the Bible, the theme about which every other in the whole book clusters, is the redemption plan, the restoration in the human soul of the image of God. From the first intimation of hope in the sentence pronounced in Eden to that last glorious promise of the Revelation, They shall see his face; and his name shall be in their foreheads (Revelation 22:4), the burden of every book and every passage of the Bible is the unfolding of this wondrous theme—man's uplifting—the power of God, which giveth us the victory through our Lord Jesus Christ. He who grasps this thought has before him an infinite field for study. He has the key that will unlock to him the whole treasure house of God's Word.... The creative energy that called the worlds into existence is in the Word of God. This word imparts power; it begets life. Every command is a promise; accepted by the will, received into the soul, it brings with it the life of the Infinite One. It transforms the nature, and re-creates the soul in the image of God. The life thus imparted is in like manner sustained. By every word that proceedeth out of the mouth of God shall man live. He may dwell in this world in the atmosphere of heaven, imparting to earth's sorrowing and tempted ones thoughts of hope and longings for holiness; himself coming closer and still closer into fellowship with the Unseen; like him of old who walked with God, drawing nearer and nearer the threshold of the eternal world, until the portals shall open, and he shall enter there. He will find himself no stranger. The voices that will greet him are the voices of the holy ones, who, unseen, were on earth his companions—voices that here he learned to distinguish and to love. He who through the Word of God has lived in fellowship with heaven will find himself at home in heaven's companionship (Education pp. 125-127)

A PLAN THAT HAS NEVER BEEN IMPROVED ON: In God's plan for Isreal every family had a home upon the land, with sufficient ground for tilling. Thus were provided both the income and the incentive for a useful, industrious, and self-supporting life. And no devising of man has ever improved upon that plan. (Ministry of Healing pp. 183-184)

GARDEN OF EDEN RESTORED IN THE EARTH MADE NEW: The Garden of Eden remained upon the earth long after man had become an outcast from its pleasant paths. The fallen race were long permitted to gaze upon the home of innocence, their entrance barred only by the watching angels. At the cherubim-guarded gate of Paradise the divine glory was revealed. Hither came Adam and his sons to worship God. Here they renewed their vows of obedience to that law the transgression of which had banished them from Eden. When the tide of iniquity overspread the world, and the wickedness of men determined their destruction by a flood of waters, the hand that had planted Eden withdrew it from the earth. But in the final restitution, when there shall be a new heaven and a new earth (Revelation 21:1), it is to be restored more gloriously adorned than at the beginning. Then they that have kept God's commandments shall breathe in immortal vigor beneath the tree of life; and through unending ages the inhabitants of sinless worlds shall behold, in that garden of delight, a sample of the perfect work of God's creation, untouched by the curse of sin—a sample of what the whole earth would have become, had man but fulfilled the Creator's glorious plan. Adam is reinstated in his first dominion. Transported with joy, he holds the trees that were once his delight—the very trees whose fruit he himself had gathered in the days of his innocence and joy. He sees the vines that his own hands have trained, the very flowers that he once loved to care for. His mind grasps the reality of the scene; he comprehends that this is indeed Eden restored. Restored to the tree of life in the long-lost Eden, the redeemed will grow up to the full stature of the race in its primeval glory. The last lingering traces of the curse of sin will be removed, and Christ's faithful ones will appear in the beauty of the Lord our God (Psalm 90:17), in mind and soul and body reflecting the perfect image of their Lord. Oh, wonderful redemption! long talked of, long hoped for, contemplated with eager anticipation, but never fully understood. (God's Amazing Grace p. 360)

WE MUST DECIDE IF WE SHALL HAVE A PLACE IN THE NEW EARTH—A TITLE TO ABRAHAM'S FARM: Many of us fail to improve our privileges. We make a few feeble efforts to do right, and then go back to our old life of sin. If we ever enter the kingdom of God, we must enter with perfect characters, not having spot, or wrinkle, or any such thing. Satan works with increased activity as we near the close of time. He lays his snares, unperceived by us, that he may take possession of our minds. In every way he tries to eclipse the glory of God from the soul. It rests with us to decide whether he shall control our hearts and minds, or whether we shall have a place in the new earth, a title to Abraham's farm. (Messages to Young People p. 105)

— (2): Advantages of Country Living for Child-rearing

CHILDREN TO BE BROUGHT INTO DIRECT CONTACT WITH THE WORKS OF GOD IN NATURE: The Lord desires His people to move into the country, where they can settle on the land, and raise their own fruit and vegetables, and where their children can be brought in direct contact with the works of God in nature. (Selected Messages Vol. 2 p. 357-358)

CHILDREN ESPECIALLY NEED THE ADVANTAGES GOD'S PLAN PROVIDES: True, you would not entirely free from annoyances and perplexing cares in the country; but you would there avoid many evils, and close the door against a flood of temptations which threaten to overpower the minds of your children. They need employment and variety. The sameness of their home makes them uneasy and restless…to live in the country would be very beneficial to them; an active, out-of-door life would develop health of both mind and body. They should have a garden to cultivate, where they might find both amusement and useful employment. The training of plants and flowers tends to the improvement of taste and judgment, while an acquaintance with God's useful and beautiful creations has a refining and ennobling influence upon the mind, referring it to the Maker and Master of all. (Testimonies to the Church Vol. 4 p. 136)

A PLAN THAT MAKES IT EASIER TO RAISE HAPPY AND USEFUL CHILDREN: It is God's design that we should love the beautiful in nature. He made a garden for our first parents, and there planted with His own divine hand the trees for usefulness and ornament, and the beautiful vines bearing fruit, and the lovely flowers of every variety and color. This was for the pleasure and happiness of man. If parents would more closely follow the example of their Creator in this respect, I believe they would have less trouble in bringing up their children to usefulness and happiness. If parents would encourage their children to love the beauties of nature, they would throw about them a safeguard to preserve them from iniquity prevailing among the youth. (*The Health Reformer*, March, 1871)

HOME OWNERSHIP ENCOURAGED: Fathers and mothers who possess a piece of land and a comfortable home are kings and queens. (Fundamentals of Christian Education p. 327)

LAND NEEDED FOR CULTIVATION: The time has come when, as God opens the way, families should move out of the cities. The children should be taken into the country. The parents should get as suitable a place as their means will allow. Though the dwelling may be small, yet there should be land in connection with it that may be cultivated. (Adventist Home p. 139)

A SMALL PIECE OF LAND IN THE COUNTRY: Educate our people to get out of the cities into the country, where they can obtain a small piece of land and make a home for themselves and their children. (Adventist Home p. 373)

COUNSEL TO MOVE TO THE COUNTRY: Get out of the country as soon as possible, and purchase a little piece of land, where you can have a garden, where your children can watch the flowers growing, and learn from them lessons of simplicity and purity. (Selected Messages Vol. 2 p. 356)

COUNTRY UPBRINGING OF ELISHA: The early years of the prophet Elisha were passed in the quietude of country life, under the teaching of God and nature and the discipline of useful work. In a time of almost universal apostasy his father's household were among the number who had not bowed the knee to Baal. Theirs was a home where God was honored and where faithfulness to duty was the rule of daily life. (Reflecting Christ p. 336)

BENEFITS OF HOME OWNERSHIP: The sense of being owners of their own homes would inspire them with a strong desire for improvement. They would soon acquire skill in planning and devising for themselves; their children would be educated to habits of industry and economy, and the intellect would be greatly strengthened. (Adventist Home p. 373)

— (3): Dangers to Children from City Living

DANGERS TO CHILDREN FROM CITY LIVING: Life in the cities is false and artificial. The intense passion for money getting, the whirl of excitement and pleasure seeking, the thirst for display, the luxury and extravagance—all are forces that, with the grat masses of mankind, are turning the mind from life's true purpose. Upon the youth they have almost irresistible power. (Adventist Home p. 135)

THE DEMORALIZING INFLUENCE OF RUSH HURRY AND NOISE: It is not God's will that His people shall settle in the cities, where there is constant turmoil and confusion. Their children should be spared this, for the whole system is demoralized by the hurry and rush and noise. (Adventist Home p. 136)

THE MISTAKES OF LOT BEING REPEATED TODAY: When Lot entered Sodom he fully intended to keep himself free from iniquity and to command his household after him. But he signally failed. The corrupting influences about him had an effect upon his own faith, and his children's connection with the inhabitants of Sodom boung up his interest in a measure with theirs. The result is before us. Many are still making a similar mistake. In selecting a home they look more to the temporal advantages they may gain than to the moral and social influences that will surround themselves and their families. They choose a beautiful and fertile country, or remove to some flourishing city, in the hope of securing greater prosperity; but their children are surrounded by temptation, and too often they form associations that are unfavorable to the development of piety and the formation of a right character. The atmosphere of lax morality, or unbelief, of indifference to religious things, has a tendency to counteract the influence of the parents. Examples of rebellion against parental and divine authority are ever before the youth; many form attachments with infidels and unbelievers, and cast in their lot with the enemies of God. (Patriarchs and Prophets p. 168-169)

LARGE CITIES IN PARTICULAR NOT FOR CHILDREN: Let it be your study to select and make your homes as far from Sodom and Gomorrah as you can. Keep out of the large cities. If possible make your homes in the quiet retirement of the country, even if you can never become wealthy by so doing. Locate where there is the best influence. (Adventist Home p. 139)

— (4): Saving Enough for a Home

A YOUNG MAN ADVISED TO ECONOMIZE AND INVEST: It is certain you have not economized in everything or you would now have something to show as the result of that wise economy which is praiseworthy in any young man. To carefully reserve a portion of each week's wages and lay by a certain sum every week which is not to be touched, should be your rule.... Diligence in business, abstinence from pleasure, even privation, so long as health is not endangered, should be cheerfully maintained by a young man in your circumstances, and you would have a little competency untouched should you become sick, that the charities of others would not be your dependence. You have needlessly expended much means which now might be on interest, and you be having some returns....You might have had, even from your limited wages, means in reserve for any demand. It might have been invested in a lot of land which would be increasing in value. But for a young man to live up to the last dollar he earns shows a great lack of calculation and discernment. (Selected Messages Vol. 2 p. 330)

EXQUISITE HOUSING NOT NEEDED: An expensive dwelling, elaborate furnishings, display, luxury, and ease, do not furnish the conditions essential to a happy, useful life. (Ministry of Healing p. 365)

EXTRAVAGANCE TO BE AVOIDED: There should be no extravagance in building fine homes, in buying costly furniture…but in everything let us think of the souls for whom Christ has died…Let us save every dollar that can be saved, that the matchless charms of Christ may be presented before the souls of the perishing. (Counsels on Stewardship p. 298)

FURTHER ADVICE ABOUT BEING ECONOMICAL ENOUGH TO SECURE PROPERTY: Brother and Sister B have not learned the lesson of economy…They would use all as they pass along, were it ever so much. They would enjoy as they go and then, when affliction draws upon them, would be wholly unprepared…Had Brother and Sister B been economical managers, denying themselves, they could ere this have had a home of their own and besides this have had means to draw upon in case of adversity. (Adventist Home p. 395)

— (5): Prudence and Wisdom Needed—Rash Moves Not to Be Made

WEIGH EVERYTHING CAREFULLY: Not a move should be made but that movement and all that it portends are carefully considered—everything weighed. (Selected Messages Vol. 2 p. 362)

MOVE IN AN ORDERLY FASHION, UNDER THE GUIDANCE OF GOD: Let there be nothing done in a disorderly manner that there shall be a great loss or sacrifice made upon property because of ardent, impulsive speeches which stir up an enthusiasm which is not after the order of God, that a victory that was essential to be gained, shall, for lack of level-headed moderation and proper contemplation and sound principles and purposes, be turned into a defeat. Let there be wise generalship in this matter, and all move under the guidance of a wise, unseen Counselor, which is God. (Selected Messages Vol. 2 p. 363)

Part 2: Powerful Testimony of Beautiful, Well-kept Grounds and Farms

THE WITNESS OF A WELL-KEPT HOME AND YARD TO CHILDREN AND COMMUNITY: Parents are under obligation to God to make their surroundings such as will correspond to the truth they profess. They can then give correct lessons to their children, and the children will learn to associate the home below with the home above. The family here must, as far as possible, be a model of the one in heaven. (Adventist Home p. 146)

A SOFTENING, RELATIONSHIP-STRENGTHENING INFLUENCE: "Nearly all dwellers in the country, however poor, could have about their homes a bit of grassy lawn, a few shade trees, flowering shrubbery, or fragrant blossoms. And far more than any artificial adorning will they minister to the happiness of the household. They will bring into the home life a softening, refining influence, strengthening the love of nature, and drawing the members of the household nearer to one another and nearer to God. (Ministry of Healing p. 370)

INCREASING REVERENCE FOR OUR CREATOR: Let the lessons of God's beautiful flowers soften, refine, and elevate your natures, and attract your hearts from the loveliness of nature to nature's God, and increase your love and reverence for your Creator. (The Health Reformer, June, 1871)

GOD DELIGHTS WHEN WE ENJOY NATURE: God loves the beautiful. He has clothed the earth and the heavens with beauty, and with a Father's joy He watches the delight of His children in the things that He has made. He desires us to surround our homes with the beauty of natural things. (Ministry of Healing p. 370)

IMPRESSIONS MADE ON VISITORS BY A PLEASANT HOME AND YARD: The proprietors of this pleasant home were in such circumstances that they might have furnished and embellished their residence expensively, but they had wisely chosen comfort rather than display. There was nothing in the house considered too good for general use, and the curtains and blinds were not kept closed to keep the carpets from fading and the furniture from tarnishing. The God- given sunlight and air had free ingress, with the fragrance of the flowers in the garden. The family were, of course, in keeping with the home; they were cheerful and entertaining, doing everything needful for our comfort, without oppressing us with so much attention as to make us fear that we were causing extra trouble. We felt that here was a place of rest. This was a home in the fullest sense of the word. (Adventist Home p. 153)

STUDENT ENERGY AND ENTHUSIASM SHOULD BE ENLISTED IN BEAUTIFYING SCHOOL GROUNDS: No recreation helpful only to themselves will prove so great a blessing to the children and youth as that which makes them helpful to others. Naturally enthusiastic and impressible, the young are quick to respond to suggestion. In planning for the culture of plants, let the teacher seek to awaken an interest in beautifying the school grounds and the schoolroom. A double benefit will result. That which the pupils seek to beautify they will be unwilling to have marred or defaced. A refined taste, a love of order, and a habit of care-taking will be encouraged; and the spirit of fellowship and co-operation developed will prove to the pupils a lifelong blessing. (Education p. 212)

A LARGE PLOT OF LAND DEDICATED TO FLOWERS AND ORNAMENTAL TREES: We have a large space of land devoted to ornamental trees and flowers. I have scoured the country for different plants and I have a large bush of lemon verbena honeysuckle. We have a large variety of roses, dahlias, gladioli, geraniums, pinks, pansies, and evergreens. This must be a sample settlement, to tell what can be raised here. (Letter 162, 1896, pp. 2-3)

THE FARM SHOULD BEAR TESTIMONY—TO ANGELS AND MEN—THAT THE WORK IS OF GOD: In regard to this school here at Huntsville, I wish to say that for the past two or three years I have been receiving instruction in regard to it— what it should be, and what those who come here as students are to become. All that is done by those connected with this school, whether they be white or black, is to be done with the realization that this is the Lord's institution, in which the students are to be taught how to cultivate the land, and how to labor for the uplifting of their own people. They are to work with such earnestness and perseverance that the farm will bear testimony, to the world, to angels, and to men, to the fidelity with which this donation of land has been cared for. This is the Lord's land, and it is to bear fruit to His glory. Those who attend this school, to be taught in right lines, on the farm or in the school, are to live in close connection with God. (Manuscript Release 6 p. 211)

SCHOOLS, FARMS, AND BUILDINGS SHOULD BE KEPT UP IN HARMONY WITH THEIR HIGH AND SACRED WORK: The Huntsville School has been presented to me as being in a very desirable location. It would be difficult to secure another location as promising as the school farm now secured. The buildings and everything connected with the work there should be in harmony with the high and sacred work to be done there. Let there be nothing unsightly connected with the buildings or about the farm, nothing that would indicate slackness. (A Place Called Oakwood p. 118)

NEGLECTED FARMS LOWER THE INFLUENCE OF A SCHOOL: If the land is well cared for, it will produce abundantly. Let the teachers go out, taking with them small companies of students, and teach these students how properly to work the soil. Let all those connected with the school study to see how they may improve the property. Teach the students to keep the gardens free from weeds. Let each one see that his room is clean and presentable. Let the care and cultivation of the land of the Huntsville School show to unbelievers that Seventh-day Adventists are reliable and that their influence is of value in the community. The sight of a farm, unproductive because of neglect, has a tendency to belittle the influence of the school. (A Place Called Oakwood p. 119)

THE EARTH IS A PRECIOUS TREASURE, LENT TO US IN TRUST: The Lord would have us treat the earth as a precious treasure, lent us in trust. (Letter 47a, 1895, pp. 5-8)

LAND IS GOD'S PROPERTY TO BE USED FOR HIS GLORY: We hope that the present sickly appearance may give place to healthful conditions. Careful cultivation will bring good returns, and the sad lack now seen may be overcome by the exercise of intelligence in determining what must be done. Let us remember that the land is God's property to be worked energetically to His glory. The trees and grains and vegetables will yield their fruit in proportion to the labor that is put forth in their care. (A Place Called Oakwood p. 119)

Part 3: Farmers Highly Effective as Missionaries

FARMERS ARE JUST AS TRULY IN THEIR LOT AND PLACE AS MINISTERS OF THE GOSPEL: In the laws which God gave for the cultivation of the soil, He was giving the people opportunity to overcome their selfishness and become heavenly-minded. Canaan would be to them as Eden if they obeyed the Word of the Lord. Through them the Lord designed to teach all the nations of the world how to cultivate the soil so that it would yield healthy fruit, free from disease. The earth is the Lord's vineyard, and is to be treated according to His plan. Those who cultivated the soil were to realize that they were doing God service. They were as truly in their lot and place as were the men appointed to minister in the priesthood and in work connected with the tabernacle. (Bible Commentary Vol. 1 p. 1112)

FARMERS (AMONG OTHER PRACTICAL PROFESSIONS) MAY BE SUCCESSFUL MISSIONARIES FOR GOD: You may be living, acceptable missionaries for God, and yet be mechanics, merchants, and farmers. (Review and Herald, March 29, 1870)

FRUIT-GROWERS, FARMERS, AND MERCHANTS CAN SAVE MANY SOULS THAT MINISTERS WILL NEVER REACH: A great work is going silently on through the distribution of our publications; but what a great amount of good might be done if some of our brethren and sisters from America would come to these colonies, as fruit-growers, farmers, or merchants, and in the fear and love of God, would seek to win souls to the truth. If such families were consecrated to God, he would use them as his agents. Ministers have their place and their work, but there are scores that the minister cannot reach, who might be reached by families who could visit with the people and impress upon them the truth for these last days. In their domestic or business relations they could come in contact with a class who are inaccessible to the minister, and they could open to them the treasures of the truth, and impart to them a knowledge of salvation. There is altogether too little done in this line of missionary work; for the field is large, and many workers could labor with success in this line of effort. If those who have received a knowledge of the truth had realized the necessity of studying the

Scriptures for themselves, if they had felt the weight of responsibility that rests upon them, as faithful stewards of the grace of God, they would have brought light to many who sit in darkness, and what a harvest of souls would have been gathered for the Master. If each one realized his accountability to God for his personal influence, he would in no case be an idler, but would cultivate his ability, and train every power that he might serve him who has purchased him with his own blood. (Review and Herald, February 14, 1893)

THINGS OF EARTH MORE CLOSELY CONNECTED WITH HEAVEN THAN PEOPLE THINK: The things of earth are more closely connected with heaven, and are more directly under the supervision of Christ, than many realize. All right inventions and improvements have their source in him who is wonderful in counsel and excellent in working. The skilful touch of the physician's hand, his power over nerve and muscle, his knowledge of the delicate mechanism of the body, is the wisdom of divine power to be used in behalf of the suffering. The skill with which the carpenter uses his tools, the strength with which the blacksmith makes the anvil ring, come from God. Whatever we do, wherever we are placed, he desires to control our minds, that we may do perfect work. Christianity and business, rightly understood, are not two separate things; they are one. Bible religion is to be brought into all that we do and say. Human and divine agencies are to combine in temporal as well as in spiritual achievements. They are to be united in all human pursuits, in mechanical and agricultural labors, in mercantile and scientific enterprises. (Review and Herald, October 3, 1912)

GOD WILL BE GLORIFIED BY AGRICULTURAL MISSIONARIES WHO SHARE THEIR KNOWLEDGE WITH OTHERS: God would be glorified if men from other countries who have acquired an intelligent knowledge of agriculture, would come to this land, and by precept and example teach the people how to cultivate the soil, that it may yield rich treasures. Men are wanted to educate others how to plow, and how to use the implements of agriculture. Who will be missionaries to do this work, to teach proper methods to the youth, and to all who feel willing and humble enough to learn? If any do not want you to give them improved ideas, let the lessons be given silently, showing what can be done in setting out orchards and planting corn; let the harvest be eloquent in favor of right methods of labor. Drop a word to your neighbors when you can, keep up the culture of your own land, and that will educate. (Special Testimonies on Education p. 101)

RELIGION TO BE WOVEN IN ALL BUSINESS PURSUITS—INCLUDING AGRICULTURE: Religion and business are not two separate things; they are one. Bible religion is to be interwoven with all we do or say. Divine and human agencies are to combine in temporal as well as in spiritual achievements. They are to be united in all human pursuits, in mechanical and agricultural labors, in mercantile and scientific enterprises. There must be co-operation in everything embraced in Christian activity. (Christ's Object Lessons p. 349)

SHOWING PEOPLE HOW TO CULTIVATE THEIR LAND IS PRACTICAL MISSIONARY WORK: I did not expect to write you in this way, but these particulars we want you to have that you may understand what we are doing. We intend to give the people practical lessons upon the improvement of the land, and thus induce them to cultivate their land, now lying idle. If we accomplish this, we shall have done good missionary work. (Manuscript Release p. 894)

SHOWING THE POOR HOW TO FARM A FAR BETTER WAY TO HELP THAN GIVING THEM MONEY: The question was asked of Moses, Can the Lord spread a table in the wilderness? The question may be asked, Will this land at Dora Creek produce as abundantly as Sister White believes that it will? Time will tell. We must test the matter before we can speak assuredly, but we are willing to risk much, provided we can place the supervision of this enterprise under an understanding America farmer. We do want to demonstrate what can be done with the land when it is properly worked. When once this is done, we shall be able to help the poor who live in Australia in a far better way than by giving them money as we have had to do in the past. (Manuscript Release 8 p. 143)

MEDICAL MISSIONARY TRAINING WITH A KNOWLEDGE OF THE TRADES ONE OF THE GRANDEST OBJECTS FOR WHICH A SCHOOL CAN BE ESTABLISHED: The school at Madison not only educates in a knowledge of the Scriptures, but it gives a practical training that fits the student to go forth as a self-supporting missionary to the field to which he is called. In his student days he is taught how to build, simply and substantially, how to cultivate the land and care for the injured. This training for medical-missionary work is one of the grandest objects for which any school can be established… (Manuscript Release 5 p. 280)

WITNESSING THROUGH DEMONSTRATION OF SOUND FARMING PRINCIPLES: Let proper methods be taught to all who are willing to learn. If any do not wish you to speak to them of advanced ideas, let the lessons be given silently. Keep up the culture of your own land. Drop a word to your neighbors when you can, and let the harvest be eloquent in favor of right methods. Demonstrate what can be done with the land when properly worked. (Ministry of Healing p. 193)

WHEN WE OBEY GOD'S LAWS, THE LAND WE CULTIVATE CAN BECOME AN OBJECT LESSON OF SPIRITUAL TRUTH: Through disobedience to God, Adam and Eve had lost Eden, and because of sin the whole earth was cursed. But if God's people followed His instruction, their land would be restored to fertility and beauty. God Himself gave them directions in regard to the culture of the soil, and they were to co-operate with Him in its restoration. Thus the whole land, under God's control, would become an object lesson of spiritual truth. As in obedience to His natural laws the earth should produce its treasures, so in obedience to His moral law the hearts of the people were to reflect the attributes of His character. Even the heathen would recognize the superiority of those who served and worshiped the living God. (Christ's Object Lessons p. 289)

THE WITNESS OF AVONDALE FARM DURING A DROUGHT: A member of parliament who came to Cooranbong occasionally, and who had purchased the house in which we first lived in Cooranbong, visited our garden and orchard, and was greatly pleased with it. Several times we filled a large basket with fruit and took it to him and his wife at their home, and they were profuse in their thanks. After this they would always recognize us on the cars, and speak of the great treat they had had in the fruit from our orchard. When they would visit us at our farm, they were always at liberty to eat all they wanted from the garden, and usually carried away a basket of fruit to their home. These favors brought us returns in several ways. Mention was made in the papers of the work being done by the students on the Avondale estate. And years afterward, when the terrible drought came, and the cattle were dying for want of pasture and food, the papers spoke of the wonderful exception to the drought to be found on the Avondale tract of land. They compared it to an oasis in the desert. Our crops were not cut off, and the farm flourished remarkably, notwithstanding the lack of rain. (Manuscript Release 15 p. 55)

Part 4: God's Economic Plan for Helping the Poor

THE SELFISHNESS, PRIDE AND OPPRESSION OF THE FEUDAL SYSTEM WOULD NEVER HAVE HAPPENED IF GOD'S PLAN FOR DIVISION OF LAND HAD BEEN ENACTED: If the laws given by God had continued to be carried out, how different would be the present condition of the world, morally, spiritually, and temporally. Selfishness and self-importance would not be manifested as now; but each would cherish a kind regard for the happiness and welfare of others, and such wide-spread destitution and human wretchedness as is now seen in most parts of England and Ireland would not exist. Instead of the poorer classes being kept under the iron heel of oppression by the wealthy, instead of having other men's brains to think and plan for them in temporal as well as in spiritual things, they would have some chance for independence of thought and action. (Historical Sketches p. 165)

UNDER GOD'S PLAN FOR DIVISION OF LAND, THE POOR AND UNFORTUNATE HAD AS MUCH CHANCE TO SUCCEED AS THEIR MORE FORTUNATE NEIGHBORS: By the special direction of God, the land was divided by lot. After it had been thus divided, no one was to feel at liberty, either from a love of change or a desire to make money, to trade his estate; neither was he to sell his land unless compelled to do so on account of poverty. And then whenever he or any of his kindred might desire to redeem it, the one who had purchased it must not refuse to sell it. And if the poor man had no one to redeem it for him, and was unable to do so himself, in the year of jubilee it should revert to him, and he should have the privilege of returning to his home and again enjoying it. Thus the poor and unfortunate were ever to have an equal chance with their more fortunate neighbors. (Historical Sketches p. 165)

GOOD MISSIONARY WORK INCLUDES HELPING PEOPLE IMPROVE THEIR LAND: We shall experiment on this land, and if we make a success, others will follow our example.... When right methods of cultivation are adopted, there will be far less poverty than now exists. We intend to give the people practical lessons upon the improvement of the land, and thus induce them to cultivate their land, now lying idle. If we accomplish this, we shall have done good missionary work. (Letter 42, 1895)

CHRISTIAN FARMERS TO AID AND INSTRUCT THE POOR: Christian farmers can do real missionary work in helping the poor to find homes on the land, and in teaching them how to till the soil and make it productive. Teach them how to use the implements of agriculture, how to cultivate various crops, how to plant and care for orchards. Ministry of Healing p. 193)

IT WAS PART OF GOD'S PLAN BOTH TO GIVE THE LAND REST AND TO CARE FOR THE POOR AND NEEDY: More than this, the Israelites were instructed to sow and reap their fields for six successive years; but every seventh year they were commanded to let the land rest. Whatever grew of itself was to be gathered by the poor; and what they left, the beasts of the field were to eat. This was to impress the people with the fact that it was God's land which they were permitted to possess for a time; that he was the rightful owner, the original proprietor, and that he would have special consideration made for the poor and unfortunate. This provision was made to lessen suffering, to bring some ray of hope, to flash some gleam of sunshine, into the lives of the suffering and distressed. Is any such statute regarded in England? Far from it. The Lord set needy human beings before the beasts; but this order has been reversed there, and, compared with the poor, horses, dogs, and other dumb animals are treated as princes. In some localities the poor are forbidden to step out of the path to pick the wild flowers which grow in abundance in many of the open fields. Anciently a man when hungry was permitted to enter another man's field or vineyard and eat as much as he chose. Even Christ and his disciples plucked and ate of the corn through which they passed. But how changed the order of things now! (Historical Sketches p. 165)

GOD'S CIVIL PLAN FOR ECONOMIC EQUALITY: On the tenth day of the seventh month, in the Day of Atonement, the trumpet of the jubilee was sounded. Throughout the land, wherever the Jewish people dwelt, the sound was heard, calling upon all the children of Jacob to welcome the year of release. On the great Day of Atonement satisfaction was made for the sins of Israel, and with gladness of heart the people would welcome the jubilee. As in the sabbatical year, the land was not to be sown or reaped, and all that it produced was to be regarded as the rightful property of the poor. Certain classes of Hebrew slaves—all who did not receive their liberty in the sabbatical year—were now set free. But that which especially distinguished the year of jubilee was the reversion of all landed property to the family of the original possessor. By the special direction of God the land had been divided by lot. After the division was made no one was at liberty to trade his estate. Neither was he to sell his land unless poverty compelled him to do so, and then, whenever he or any of his kindred might desire to redeem it, the purchaser must not refuse to sell it; and if unredeemed, it would revert to its first possessor or his heirs in the year of jubilee....The people were to be impressed with the fact that it was God's land which they were permitted to possess for a time; that He was the rightful owner, the original proprietor, and that He would have special consideration made for the poor and unfortunate. It was to be impressed upon the minds of all that the poor have as much right to a place in God's world as have the more wealthy. Such were the provisions made by our merciful Creator, to lessen suffering, to bring some ray of hope, to flash some gleam of sunshine, into the life of the destitute and distressed. (Patriarchs and Prophets p. 533-534)

GOD'S PLAN FOR DISTRIBUTION OF PROPERTY WOULD, IF CARRIED OUT TODAY, VASTLY IMPACT THE WORLD: Were the principles of God's laws regarding the distribution of property carried out in the world today, how different would be the condition of the people! (Education p. 42-44)

GOD'S PLAN FOR ISRAEL'S LAND SAFEGUARDED AGAINST EXTREMES IN POVERTY AND WEALTH: In apportioning the inheritance of His people, it was God's purpose to teach them, and through them the people of after generations, correct principles concerning the ownership of the land. The land of Canaan was divided among the whole people, the Levites only, as ministers of the sanctuary, being excepted. Though one might for a season dispose of his possession, he could not barter away the inheritance of his children. When able to do so, he was at liberty at any time to redeem it; debts were remitted every seventh year, and in the fiftieth, or year of jubilee, all landed property reverted to the original owner. Thus every family was secured in its possession, and a safeguard was afforded against the extremes either of wealth or of poverty. (To Be Like Jesus p. 250)

Part 5: Ellen White's Missionary Work Helping Farm Families Get Started

ELLEN WHITE INVOLVED IN THE MISSIONARY WORK OF SETTLING POOR FAMILIES ON FARMLAND IN AUSTRALIA: I shall do as I wrote you. I promised to take the school ground as my property, and I will not consider it a hard matter. I think no better missionary work could be done than to settle poor families on the land. Every family shall sign a contract that they will work the land according to the plans specified. Someone must be appointed to direct the working of the land, and under his supervision orange trees, and fruit trees of every appropriate description should be planted. Peach orchards would yield quick return. Vegetable gardens would bring forth good crops. This must be done at once. We have some six weeks yet to set things in running order, and with God's blessing on the land, we shall see what it will produce. (Manuscript Release 8 p. 143)

ELLEN WHITE INVOLVED IN SECURING LAND, BRINGING OVER MISSIONARY FARMERS AND PLANNING WHAT WOULD BE PLANTED: The more I see the school property the more I am amazed at the cheap price at which it has been purchased. When the board want to go back on this purchase, I pledge myself to secure the land. I will settle it with poor families; I will have missionary families come out from America and do the best kind of missionary work in educating the people as to how to till the soil and make it productive. I have planned what can be raised in different places. I have said, Here can be alfalfa, there can be strawberries, here can be sweet corn and common corn, and this ground will raise good potatoes, while that will raise good fruit of all kinds. So in imagination I have all the different places in flourishing condition. (Manuscript Release 13 p. 351)

ELLEN WHITE ENGAGED IN EXPERIMENTAL FARMING AND TEACHING PEOPLE HOW TO GROW AS A FORM OF MISSIONARY WORK: We have located [Avondale] here on missionary soil, and we design to teach the people all round us how to cultivate the land. They are all poor because they have left their land uncultivated. We are experimenting, and showing them what can be done in fruit raising and gardening. (Manuscript Release 7 p. 253)

TESTIMONY OF SUCCCESSFUL CROPS: Our crops were very successful. The peaches were the most beautiful in coloring, and the most delicious in flavor of any that I had tasted. (Letter 350, 1907)

ELLEN WHITE'S EXAMPLE OF TEACHING LIFE SKILLS: We did what we could to develop our land, and encouraged our neighbors to cultivate the soil, that they too might have fruits and vegetables of their own. We taught them how to prepare the soil, what to plant, and how to take care of the growing produce. They soon learned the advantages of providing for themselves in this way. (Welfare Ministry p. 328)

SUGGESTIONS ABOUT PROVIDING JOBS IN RURAL AREAS: Believers who are now living in the cities will have to move to the country, that they may save their children from ruin. Attention must be given to the establishment of industries in which these families can find employment. Those who have charge of the school-work at——— and———should see what can be done by these institutions to establish such industries, so that our people desiring to leave the cities, can obtain modest homes without a large outlay of means, and can also find employment....All that needs to be done cannot be specified till a beginning is made. Pray over the matter, and remember that God stands at the helm, that He is guiding in the work of the various enterprises. A place in which the work is conducted on right lines is an object lesson to other places. There must be no narrowness, no selfishness, in the work done. The work is to be placed on a simple, sensible basis. All are to be taught not only to claim to believe the truth, as the truth, but to exemplify the truth in the daily life. (Country Living pp. 19-20)

A SCHOOL THAT PLANNED TO EXPERIMENT AND LEAD THE WAY IN AGRICULTURE: The school land, fifteen hundred acres, was purchased for $5,500. The school has twelve acres put into orchard, I have two acres in fruit trees. We shall experiment on this land, and if we make a success, others will follow our example. Notwithstanding oranges and lemons have yielded year after year, not a new tree is planted by the settlers. Their indolence and laziness causes false witness to be borne against the land. When right methods of cultivation are adopted there will be far less poverty than now exists. (Manuscript Release p. 894)

LIGHT FROM HEAVEN ABOUT HOW TO PLANT FRUIT TREES: While we were in Australia, we adopted the plan of digging deep trenches and filling them in with dressing that would create good soil. This we did in the cultivation of tomatoes, oranges, lemons, peaches, and grapes. The man of whom we purchased our peach trees told me that he would be pleased to have me observe the way they were planted. I then asked him to let me show him how it had been represented in the night season that they should be planted. I ordered my hired man to dig a deep cavity in the ground, then put in rich dirt, then stones, then rich dirt. After this he put in layers of earth and dressing until the hole was filled. I told the nurseryman that I had planted in this way in the rocky soil in America. I invited him to visit me when these fruits should be ripe. He said to me, 'You need no lesson from me to teach you how to plant the trees.' Our crops were very successful. The peaches were the most beautiful in coloring, and the most delicious in flavor of any that I had tasted. We grew the large yellow Crawford and other varieties, grapes apricots, nectarines, and plums. (Letter 350, 1907)

Part 6: Scattered as Beacons of Light

Those who are wise shall shine like the brightness of the firmament, and those who turn many to righteousness like the stars forever and ever. Daniel 12:3

MISSIONARY FARMERS (AMONG OTHER PROFESSIONS) NEEDED TO SETTLE IN PLACES WHERE CHRIST IS NOT KNOWN: Missionary families are needed to settle in the waste places. Let farmers, financiers, builders, and those who are skilled in various arts and crafts, go to neglected fields, to improve the land, to establish industries, to prepare humble homes for themselves, and to help their neighbors. (Ministry of Healing p. 194)

MISSIONARY FAMILIES NEEDED TO MOVE TO THE WASTE PLACES: Missionary families are needed to settle in the waste places. Let farmers, financiers, builders, and those who are skilled in various arts and crafts, go to neglected fields, to improve the land, to establish industries, to prepare humble homes for themselves, and to help their neighbors. (Ministry of Healing p. 194)

AVOID COLONIZING AT INSTITUTIONAL CENTERS: Many families, who, for the purpose of educating their children, move to places where our large schools are established, would do better service for the Master by remaining where they are. They should encourage the church of which they are members to establish a church school where the children within their borders could receive an all-round, practical Christian education. It would be vastly better for their children, for themselves, and for the cause of God, if they would remain in the smaller churches, where their help is needed. (Counsels to Parents, Teachers and Students p. 173-174)

NEHEMIAH'S ARE NEEDED: We need Nehemiahs.... His energy and determination inspired the people of Jerusalem; and strength and courage took the place of feebleness and discouragement. His holy purpose, his high hope, his cheerful consecration to the work, were contagious. The people caught the enthusiasm of their leader, and in his sphere each man became a Nehemiah, and helped make stronger the hand and heart of his neighbor. Here is a lesson for ministers of the present day. (Bible Commentary Vol. 3 p. 1137)

GOD'S PEOPLE TO GO FORTH: It is not the purpose of God that His people should colonize or settle together in large communities. The disciples of Christ are His representatives upon the earth, and God designs that they shall be scattered all over the country, in the towns, cities, and villages, as lights amidst the darkness of the world. (Christian Service p. 178)

AS PILGRIMS AND STRANGERS: True missionary workers will not colonize. God's people are to be pilgrims and strangers on the earth. The investment of large sums of money in the building up of the work in one place is not in the order of God. Plants are to be made in many places. Schools and sanitariums are to be established in places where there is now nothing to represent the truth. These interests are not to be established for the purpose of making money, but for the purpose of spreading the truth. Land should be secured at a distance from the cities, where schools can be built up in which the youth can be given an education in agricultural and mechanical lines. (Testimonies to the Church Vol. 8 p. 215)

THE RESULT OF THIS GOING FORTH: Wonderful revivals will follow. Sinners will be converted, and many souls will be added to the church. When we bring our hearts into unity with Christ, and our lives into harmony with His work, the Spirit that fell on the disciples on the Day of Pentecost will fall on us. (Testimonies for the Church Vol. 8 p. 246)

THOSE WHO SHARE SACRED TRUTH—WHO PANT AFTER HOLINESS—WILL HAVE MORE AND MORE OF GRACE: In proportion as the Lord has made them the depositaries of sacred truth will be their desire that others shall receive the same blessing. And as they make known the rich treasures of God's grace, more and still more of the grace of Christ will be imparted to them. They will have the heart of a little child in its simplicity and unreserved obedience. Their souls will pant after holiness, and more and more of the treasures of truth and grace will be revealed to them to be given to the world. (Lift Him Up p. 112)

GOD CALLS FARMERS—AS WELL AS OTHER PROFESSIONS—TO REVEAL CHRIST IN THEIR WORK: To everyone who becomes a partaker of His grace the Lord appoints a work for others. Individually we are to stand in our lot and place, saying, Here am I; send me (Isaiah 6:8). Upon the minister of the Word, the missionary nurse, the Christian physician, the individual Christian whether he be merchant or farmer, professional man or mechanic—the responsibility rests upon all. It is our work to reveal to men the gospel of their salvation. Every enterprise in which we engage should be a means to this end.... (Reflecting Christ p. 231)

AN UNNATURAL THING TO KEEP OUR WITNESS SECRET: All who receive the gospel message into the heart will long to proclaim it. The heaven-born love of Christ must find expression. Those who have put on Christ will relate their experience, tracing step by step the leadings of the Holy Spirit—their hungering and thirsting for the knowledge of God and of Jesus Christ whom He has sent, the results of their searching of the Scriptures, their prayers, their soul agony, and the words of Christ to them, Thy sins be forgiven thee. It is unnatural for any to keep these things secret, and those who are filled with the love of Christ will not do so. (Lift Him Up p. 112)

SOUL SAVING WORK NOT JUST FOR MINISTERS: It is a fatal mistake to suppose that the work of soul saving depends alone upon the ministry. The humble, consecrated believer upon whom the Master of the vineyard places a burden for souls, is to be given encouragement by the men upon whom the Lord has laid larger responsibilities. Those who stand as leaders in the church of God are to realize that the Savior's commission is given to all who believe in His name. God will send forth into His vineyard many who have not been dedicated to the ministry by the laying on of hands. (The Faith I Live By p. 308)

CONSECRETED MEMBERS TO DO A LARGE WORK OUTSIDE OF THE PULPIT: Hundreds, yea, thousands, who have heard the message of salvation, are still idlers in the market place, when they might be engaged in some line of active service. To these Christ is saying, Why stand ye here all the day idle? and He adds, Go ye also into the vineyard. (Matthew 20:6-7) Why is it that many more do not respond to the call? Is it because they think themselves excused in that they do not stand in the pulpit? Let them understand that there is a large work to be done outside the pulpit, by thousands of consecrated lay members. (The Faith I Live By p. 308)

EARNEST WORK THAT INSPIRED OTHERS TO ZEAL: I came to this place, and began work on my place so earnestly that it inspired all with fresh zeal, and they have been working with a will, rejoicing that they have the privilege. We have provoked one another to zeal and good works. (Testimonies to Ministers and Gospel Workers p. 242)

GOD WAITING LONG FOR CHURCH MEMBERS TO FULFIL THE GOSPEL COMMISSION: Long has God waited for the spirit of service to take possession of the whole church, so that everyone shall be working for Him according to his ability. When the members of the church of God do their appointed work in the needy fields at home and abroad, in fulfillment of the gospel commission, the whole world will soon be warned, and the Lord Jesus will return to this earth with power and great glory. (The Faith I Live By p. 308)

HUMANS SHOULD EMULATE NATURE IN STRIVING WITHIN THEIR SPHERE: As the things of nature show their appreciation of the Master Worker by doing their best to beautify the earth and to represent God's perfection, so human beings should strive in their sphere to represent God's perfection, allowing Him to work out through them His purposes of justice, mercy, and goodness. (Child Guidance 54)

WE MAY HELP THE HEAVENLY SHEPHERD FIND SHEEP WHO STRAY FROM THE FOLD: The heavenly Shepherd knows where to find the lambs that are straying from the fold. He will gather them in. He calls upon ministers and lay members to arouse to their responsibility, and unite with Him in this work. It is the special duty of Christians to seek and save the lost. Ministers and laymen are to encourage and help those who, sorely beset by temptation, know not which way to turn. My brother, through the grace of God you may become one who is able to bring back to the fold the wandering ones. (This Day With God p. 67)

A LIFE FULL OF GRACIOUS OPPORTUNITIES: This life is full of gracious opportunities, which you can improve in the exercise of your God-given abilities to bless others, and in so doing bless yourself, without considering self in the matter. Trivial circumstances oftentimes prove a decided blessing to the one who acts from principle and has formed the habit of doing right because it is right. Seek for a perfect character, and let all you do, whether seen and appreciated by human eyes or not, be done with an eye single to God's glory, because you belong to God and He has redeemed you at the price of His own life. Be faithful in the least as well as in the greatest; learn to speak the truth, to act all times the truth. Let the heart be fully submitted to God. If controlled by His grace, you will do little deeds of kindness, take up the duties lying next to you, and bring all the sunshine into your life and character that it is possible to bring, scattering the gifts of love and blessing along the pathway of life. Your works will be far-reaching as eternity. Your lifework will be seen in heaven, and there it will live, through ceaseless ages, because it is found precious in the sight of God. (My Life Today p. 219)

THIS WORLD IS NOT OUR HOME: God promised to Abraham, and his seed after him, that they should have possessions and lands, and yet they were only strangers and sojourners. The inheritance and lands that are to be given not only to Abraham but to the children of Abraham will not be until after this earth is purified. Abraham will then receive the title to his farm, his possessions; and the children of Abraham will have a title to their possessions. Every one of us should constantly bear in mind that this earth is not our dwelling place, but that we are to have an inheritance in the earth made new. The destruction of Sodom and Gomorrah symbolizes to us how this world will be destroyed by fire. It is not safe for any one of us to build our hopes in this life. We want first to seek the kingdom of God and His righteousness.... (Christ Triumphant p. 80)

OUR HEARTS NOT TO BE ON EARTHLY TREASURES—FARMING OR OTHERWISE: I have been shown that there was not that being done which God has a right to expect of you in New York State to advance his cause and push forward the work, in wisely investing his entrusted talents. All the money is the Lord's. Why do you withhold from God that which is his own? There is not one hundredth part being done that ought to be done in your State. There is so great lack of faith and corresponding works that God cannot do much for you. The narrow faith, the narrow plans, are the limiting and binding about of the work. God will work for us just in accordance with our faith. At the slow rate our people in many States are working, it would take a temporal millennium to warn the world. The angels are holding the four winds that they should not blow until the world is warned, until a people has decided for the truth, the honest of heart have been convicted and converted. Their power, their influence, and their means will then flow in the missionary channel. This is putting out the money to the exchangers, that when the Master shall come his stewards may present the talents doubled in the ingathering of souls to Jesus Christ. But the wealthy farmers are some of them acting as if in the day of God the Lord only would require of them to present to him enriched, improved farms, building added to building, and they say, Here Lord are thy talents; behold, I have gained all this possession. If the acres of their farms were so many precious souls saved to Jesus Christ, if their buildings were so many souls to be presented to the Master, then he could say to these men, Well done, good and faithful servant. But you cannot take these improved farms, or these buildings into heaven. The fires of the last days will consume them. If you invest and bury your talents of means in these earthly treasures, your heart is on them, your anxiety is for them, your persevering labor is for them, your tact, your skill is cultivated to serve earthly, worldly possessions, and are not directed or employed upon heavenly things. And you come to look upon means invested for larger plans in extending the work as so much means lost which bring no returns. This is all a mistake, because the earthly is exalted above the eternal. While the heart is on earthly treasures it can only estimate such; it cannot appreciate the heavenly treasure. It is fully occupied just as the Devil wants it should be; and the eternal is eclipsed by the earthly. (Pamphlet 39, "An Important Testimony to our Brethren and Sisters in New York")

OUR TIME IS A TALENT: To every man is committed individual gifts, termed talents. Some regard these talents as being limited to certain men who possess superior mental endowments and genius. But God has not restricted the bestowal of His talents to a favored few. To everyone is committed some special endowment, for which he will be held responsible by the Lord. Time, reason, means, strength, mental powers, tenderness of heart—all are gifts from God, entrusted to be used in the great work of blessing humanity. (God's Amazing Grace p. 64)

ENDEAVOR TO SAVE SOULS WHEREVER WE WORK: There is a great work to be done in our world. Men and women are to be converted, not by the gift of tongues nor by the working of miracles, but by the preaching of Christ crucified. Why delay the effort to make the world better? Why wait for some wonderful thing to be done, some costly apparatus to be provided? However humble your sphere, however lowly your work, if you labor in harmony with the teachings of the Savior, He will reveal Himself through you, and your influence will draw souls to Him. He will honor the meek and lowly ones, who seek earnestly to do service for Him. Into all that we do, whether our work be in the shop, on the farm, or in the office, we are to bring the endeavor to save souls. (Reflecting Christ p. 256)

THE SCHOOL IN COORANBONG TO BE A CENTER OF LIGHT, TRAINING PEOPLE TO WORK UNIMPROVED LANDS: In the school that is started here in Cooranbong, we look to see real success in agricultural lines, combined with a study of the sciences. We mean for this place to be a center, from which shall irradiate light, precious advanced knowledge that shall result in the working of unimproved lands, so that hills and valleys shall blossom like the rose. For both children and men, labor combined with mental taxation will give the right kind of all-round education. The cultivation of the mind will bring tact and fresh incentives to the cultivation of the soil. (Testimonies to Ministers and Gospel Workers p. 244)

A TRULY IMPRESSIVE VISION: In the visions of the night a very impressive scene passed before me. I saw an immense ball of fire fall among some beautiful mansions, causing their instant destruction. I heard someone say: We knew that the judgments of God were coming upon the earth, but we did not know that they would come so soon. Others, with agonized voices, said: You knew! Why then did you not tell us? We did not know. On every side I heard similar words of reproach spoken. In great distress I awoke. I went to sleep again, and I seemed to be in a large gathering. One of authority was addressing the company, before whom was spread out a map of the world. He said that the map pictured God's vineyard, which must be cultivated. As light from heaven shone upon anyone, that one was to reflect the light to others. Lights were to be kindled in many places, and from these lights still other lights were to be kindled. The words were repeated: Ye are the salt of the earth: but if the salt have lost his savor, wherewith shall it be salted? it is thenceforth good for nothing, but to be cast out, and to be trodden underfoot of men. Ye are the light of the world. A city that is set on an hill cannot be hid. Neither do men light a candle, and put it under a bushel, but on a candlestick; and it giveth light unto all that are in the house. Let your light so shine before men, that they may see your good works, and glorify your Father which is in heaven. Matthew 5:13-16. I saw jets of light shining from cities and villages, and from the high places and the low places of the earth. God's word was obeyed, and as a result there were memorials for Him in every city and village. His truth was proclaimed throughout the world. Then this map was removed and another put in its place. On it light was shining from a few places only. The rest of the world was in darkness, with only a glimmer of light here and there. Our Instructor said: This darkness is the result of men's following their own course. They have cherished hereditary and cultivated tendencies to evil. They have made questioning and faultfinding and accusing the chief business of their lives. Their hearts are not right with God. They have hidden their light under a bushel. If every soldier of Christ had done his duty, if every watchman on the walls of Zion had given the trumpet a certain sound, the world might ere this have heard the message of warning. But the work is years behind. While men have slept, Satan has stolen a march upon us. Putting our trust in God, we are to move steadily forward, doing His work with unselfishness, in humble dependence upon Him, committing ourselves and our present and future to His wise providence, holding the beginning of our confidence firm unto the end, remembering that it is not because of our worthiness that we receive the blessings of heaven, but because of the worthiness of Christ, and our acceptance, through faith in Him, of God's abounding grace. (Testimonies to the Church Vol. 9 pp. 28-29)

A PROMISE FOR TOWNS AND CITIES NEGLECTED: In that day shall the deaf hear the words of the book, and the eyes of the blind shall see out of obscurity, and out of darkness. The meek also shall increase their joy in the Lord, and the poor among men shall rejoice in the Holy One of Israel. (Isaiah 29:18-19)

Then the trees of the field shall yield their fruit, and the earth shall yield her increase. They shall be safe in their land; and they shall know that I am the Lord, when I have broken the bands of their yoke and delivered them from the hand of those who enslaved them. Ezekiel 34:27, NKJV

Chapter #4: Blessings to be Obtained through Involvement in Farming

Blessing #1: Character Traits Developed

— (1): Faithfulness in Little Things

FAITHFULNESS REQUIRED IN ALL AREAS OF SCHOOL-RELATED ENDEAVOR—INCLUDING AGRICULTURE: Let none of our schools be conducted in a cheap, careless manner. He that is faithful in that which is least will be faithful also in that which is greater. If little things are left uncorrected, there is danger that larger evils will be regarded indifferently. The faithful steward of the Lord's treasure will correct at once the small mistakes. Whether his duties are connected with the cultivation of the Lord's land, or with the buildings that are erected on the land, he will do every stroke well. The Lord will take notice of his faithfulness, and He will strengthen the ability to plan and execute in temporal matters. And this faithful exactitude is a special necessity where eternal interests are involved. (A Place Called Oakwood p. 121)

BEING A SUCCESSFUL FARMER REQUIRES FAITHFULNESS IN THE LITTLE DUTIES OF LIFE: I have a case now in mind of one who was presented before me in vision who neglected these little things [duties about the house—chores and little errands] and could not interest himself in small duties, seeking to lighten the work of those indoors; it was too small business. He now has a family, but he still possesses the same unwillingness to engage in these small yet important duties. The result is, great care rests upon his wife. She has to do many things, or they will be left undone; and the amount of care which comes upon her because of her husband's lack is breaking her constitution. He cannot now overcome this evil as easily as he could in his youth. He neglects the little duties and fails to keep everything up tidy and nice, therefore cannot make a successful farmer. He that is faithful in that which is least is faithful also in much:

and he that is unjust in the least is unjust also in much. (Testimonies to the Church Vol. 2 p. 309)

ELISHA, WHO WAS FAITHFUL IN THE LITTLE THINGS, LEARNED HOW TO COOPERATE WITH GOD ON HIS FATHER'S FARM: The surroundings of Elisha's home were those of wealth; but he realized that in order to obtain an all-round education, he must be a constant worker in any work that needed to be done. He had not consented to be in any respect less informed than his father's servants. He had learned how to serve first, that he might know how to lead, instruct, and command. Elisha waited contentedly, doing his work with fidelity. Day by day, through practical obedience and the divine grace in which he trusted, he obtained rectitude and strength of purpose. While doing all that he possibly could in cooperating with his father in the home firm, he was doing God's service. He was learning how to cooperate with God. (Youth Instructor, April 14, 1898)

LESSONS FROM SHRUB TRAINING ON PERFORMING LIFE'S LITTLE DUTIES: In this world we have temporal duties to perform, and in the performance of these duties we are forming characters that will either stand the test of the judgment or be weighed in the balances and found wanting. We may do the smallest duties nobly, firmly, faithfully, as if seeing the whole heavenly host looking upon us. Take a lesson from the gardener. If he wishes a plant to grow he cultivates and trims it; he gives water, he digs about its roots, plants it where the sunshine will fall upon it, and day by day he works about it; and not by violent efforts, but by acts constantly repeated, he trains the shrub until its form is perfect and its bloom is full. The grace of our Lord Jesus Christ works upon the heart and mind as an educator. The continued influence of His Spirit upon the soul trains and molds and fashions the character after the divine model. Let the youth bear in mind that a repetition of acts, forms habits, and habit, character.... Is the love of Christ a living, active agent in your soul, correcting, reforming, refining you, and purifying you from your wrong practices? There is need of cultivating every grace that Jesus through His suffering and death has brought within your reach. You are to manifest the grace that has been so richly provided for you, in the small as well as in the large concerns of life.... Great truths can be brought into little things, and religion can be carried into the little as well as into the large concerns of life. (That I May Know Him p. 157)

Chapter #4: Blessings to be Obtained through Involvement in Farming

THE INFLUENCE OF A LIFE CLOSE TO NATURE—AND FAITHFULNESS IN LITTLE THINGS—IN THE CHARACTER DEVELOPMENT OF JOSEPH: From the dungeon Joseph was exalted to be ruler over all the land of Egypt. It was a position of high honor, yet it was beset with difficulty and peril. One cannot stand upon a lofty height without danger. As the tempest leaves unharmed the lowly flower of the valley, while it uproots the stately tree upon the mountaintop, so those who have maintained their integrity in humble life may be dragged down to the pit by the temptations that assail worldly success and honor. But Joseph's character bore the test alike of adversity and prosperity. The same fidelity to God was manifest when he stood in the palace of the Pharaohs as when in a prisoner's cell. He was still a stranger in a heathen land, separated from his kindred, the worshipers of God; but he fully believed that the divine hand had directed his steps, and in constant reliance upon God he faithfully discharged the duties of his position.... In his early years he had consulted duty rather than inclination; and the integrity, the simple trust, the noble nature, of the youth bore fruit in the deeds of the man. A pure and simple life had favored the vigorous development of both physical and intellectual powers. Communion with God through His works and the contemplation of the grand truths entrusted to the inheritors of faith had elevated and ennobled his spiritual nature, broadening and strengthening the mind as no other study could do. Faithful attention to duty in every station, from the lowliest to the most exalted, had been training every power for its highest service. He who lives in accordance with the Creator's will is securing to himself the truest and noblest development of character.... There are few who realize the influence of the little things of life upon the development of character. Nothing with which we have to do is really small. The varied circumstances that we meet day by day are designed to test our faithfulness and to qualify us for greater trusts. By adherence to principle in the transactions of ordinary life, the mind becomes accustomed to hold the claims of duty above those of pleasure and inclination (Patriarchs and Prophets p. 222-223)

THE FAITHFULNESS AND INTEGRITY OF ELISHA WERE LEARNED IN QUIETUDE AND SIMPLICITY. HE GREW UP ON A FARM: The attention of Elijah was attracted to Elisha, the son of Shaphat, who with the servants was plowing with twelve yoke of oxen. He was educator, director, and worker. Elisha did not live in the thickly populated cities. His father was a tiller of the soil, a farmer. Far from city and court dissipation, Elisha had received his education. He had been trained in habits of simplicity, of obedience to his parents and to God. Thus in quietude and contentment he was prepared to do the humble work of cultivating the soil. But though of a meek and quiet spirit, Elisha had no changeable character. Integrity and fidelity and the love and fear of God were his. He had the characteristics of a ruler, but with it all was the meekness of one who would serve. His mind had been exercised in the little things, to be faithful in whatsoever he should do; so that if God should call him to act more directly for him, he would be prepared to hear his voice. (Youth Instructor, April 14, 1898)

NEED TO CARE FOR THE LITTLE THINGS. SLOTHFULNESS DEMORALIZES: Many persons have not been educated to care for the little things. Yet such an education is necessary to success. Those who reach a high standard must overcome the tendency to slothfulness. A tendency to neglect something that should be done at once grows into a habit of indolence. See that broken plastering, broken furniture, or broken carriages are promptly put in repair. Slothfulness in character is demoralizing. (A Place Called Oakwood p. 121)

PRIDE IN A WEED-FREE GARDEN: You may have pride in keeping out every weed. (The Health Reformer, June, 1871)

SURFACE WORK AND HAPHAZARD EFFORT REVEAL THEMSELVES IN THE HARVEST: There must be an intelligent, harmonious cooperation of the divine and human. The working of the soil is a lesson book, which if read will be of the greatest benefit to every student in our school. They may understand that surface work, haphazard half-effort, will reveal itself in the harvest to be garnered… (Manuscript Release 860)

THOSE WHO WOULD SERVE IN HIGHER POSITIONS MUST BE RELIABLE IN THE THINGS THAT ARE LEAST: The youth should bear in mind that their physical strength, their mental qualifications, and their spiritual endowments, are to be devoted to service. These are never to be misapplied, never misused, never left to rust through inaction. Elisha increased in knowledge daily. Daily he prepared to do service in any way that opened before him. He served God in the little temporal duties. He grew in knowledge and in grace. And if the student today will develop reliability and soundness of principle in the things which are least, he will reveal that he has acquired adaptability to serve God in a higher capacity. He who feels that it is of no great consequence to serve in the lesser capacity will never be trusted of God to serve in the more honored position.

He may present himself as fully competent to accomplish the duties of the higher position; but God looks deeper than the surface. A watcher is on his track, and after test and trial, there is written against him, Thou art weighed in the balances, and art found wanting. That sentence in the courts of heaven decides for eternity the destiny of the human being. (Youth Instructor, April 14, 1898)

— (2): Cooperation with God

GOD WANTS US TO WORK TOGETHER WITH HIM: In our labor we are to be workers together with God. He gives us the earth and its treasures; but we must adapt them to our use and comfort. He causes the trees to grow; but we prepare the timber and build the house. He has hidden in the earth the gold and silver, the iron and coal; but it is only through toil that we can obtain them.... (To Be Like Jesus p. 233)

IN BOTH THE SPIRITUAL AND AGRICULTURAL REALM, WE MUST WORK HAND-IN-HAND WITH GOD: We need to understand that individually we are in copartnership with God. Work out your own salvation with fear and trembling, He admonishes us, and adds, For it is God which worketh in you both to will and to do of his good pleasure (Philippians 2:12, 13). Here is the cooperation of the divine with human agencies....The former and the latter rains are needed. We are labourers together with God (1 Corinthians 3:9). The Lord alone can give the precious former and latter rain. The clouds, the sunshine, the dews at night—these are heaven's most precious provisions. But all these favors graciously bestowed of Heaven will prove of little worth to those who do not appropriate them by diligent, painstaking effort on their part. Personal efforts must be put forth in agriculture. There is the plowing and replowing. Implements must be brought in and human skill must use them. The seed must be sown in its season. The laws which control seedtime and harvest must be observed, else there will be no harvest....The apostle brings in another figure: Ye are God's building (Verse 9)—an edifice to be erected. The construction of a building calls for skill in using the timber which God has caused to grow for the happiness and blessing of man. The Lord has provided the forest trees, and now man must use the trees. They must be cut down and prepared by saw, and axe, and wedge, and hammer, to be fitted for the building....Thus is presented the copartnership of the human and the divine. All the power is of God. Without me, says

Christ, ye can do nothing (John 15:5). Then how many hours is it safe for us to try to work alone? All the glory proceeds from God and should flow back in all possible ways to God, through our cooperation with God....We need to consider carefully our own spiritual interest. If we are abiding in Christ, we shall not allow ambitious business transactions, even in our service for Him, to come before the spiritual fragrance that should characterize our association with our brethren. (Manuscript 182, September 24, 1897, "Ye Are God's Husbandry")

OUR JOBS ARE NOT UNLIKE GOD'S: While God has created and constantly controls all things, He has endowed us with a power not wholly unlike His. To us has been given a degree of control over the forces of nature. As God called forth the earth in its beauty out of chaos, so we can bring order and beauty out of confusion. And though all things are now marred with evil, yet in our completed work we feel a joy akin to His when, looking on the fair earth, He pronounced it very good. (To Be Like Jesus p. 233)

— (3): Grace and Spiritual Nerve

MORE GRACE AND SPIRITUAL NERVE NEEDED TO WALK IN THE COMMON INDUSTRIES OF LIFE THAN TO ENGAGE IN OPEN MISSIONARY SERVICE: The essential lesson of contented industry in the necessary duties of life is yet to be learned by many of Christ's followers. It requires more grace, more stern discipline of character, to work for God in the capacity of mechanic, merchant, lawyer, or farmer, carrying the precepts of Christianity into the ordinary business of life, than to labor as an acknowledged missionary in the open field. It requires a strong spiritual nerve to bring religion into the workshop and the business office, sanctifying the details of everyday life, and ordering every transaction according to the standard of God's word. But this is what the Lord requires. (Review and Herald, October 3, 1912)

— (4): Industry, Religious Thrift, and Dependence on God

LESSONS OF INDUSTRY AND DEPENDENCE ON GOD LEARNED IN THE PROCESS OF FARMING: As they cultivate the soil, the students are to learn spiritual lessons. The plow must break up the fallow ground. It must lie under the rays of the sun and the purifying air. Then the seed, to all appearance dead, is to be dropped into the prepared soil. Trees are to be planted, seeds for vegetables sown. And after man has acted his part, God's miracle-working power gives life and vitality to the things placed in the soil. In this agricultural process, there are lessons to be learned. Man is not to do slothful work. He is to act the part appointed him by God. His industry is essential if he would have a harvest. (Manuscript Release 11 p. 177)

ALL WORK SHOULD BE DONE WITH RELIGIOUS THRIFT: Several acres of land should at the right time be set out to tomatoes. Young plants should be ready to be transplanted as early as possible. Such a crop would be valuable and might be used to good advantage. Let everything reveal religious thrift. (A Place Called Oakwood p. 121)

DEPENDENCE ON GOD—THE VERY FIRST LESSON TO LEARN—TAUGHT BY THE FLOWERS: The first lesson to be taught ... is the lesson of dependence upon God.... As a flower of the field has its root in the soil; as it must receive air, dew, showers, and sunshine, so must we receive from God that which ministers to the life of the soul. (Our Father Cares p. 21)

— (5): Intelligence and Innovation

INTELLIGENCE AND CONSTANT LEARNING NEEDED BY FARMERS: Farmers need far more intelligence in their work. In most cases it is their own fault if they do not see the land yielding its harvest. They should be constantly learning how to secure a variety of treasures from the earth. (Special Testimonies on Education p. 102)

INTELLIGENCE NEEDED TO FARM—WISDOM WILL BE LEARNED BY FAILURES: The cultivation of our land requires the exercise of all the brain power and tact we possess. The unworked lands around us testify to the indolence of men. We hope to arouse to action the dormant senses. We hope to see intelligent farmers, who will be rewarded for their earnest labor. The hand and head must cooperate, bringing new and sensible plans into operation in the cultivation of the soil. We have here seen the giant trees felled and uprooted, we have seen the ploughshare pressed into the earth, turning deep furrows for the planting of trees and the sowing of the seed. The students are learning what ploughing means, and that the hoe and the shovel and the rake and the harrow are all implements of honorable and profitable industry. Mistakes will often be made, but every error lies close beside truth. Wisdom will be learned by failures, and the energy that will make a beginning gives hope of success in the end. . . . (Letter 47a, 1895)

EDUCATED ABILITY NEEDED: Let the educated ability be employed in devising improved methods of work. This is what the Lord wants. There is honor in any class of work that is essential to be done. Let the law of God be made the standard of action, and it ennobles and sanctifies all labor. Faithfulness in the discharge of every duty makes the work noble, and reveals a character that God can approve. (Fundamentals of Christian Education p. 315)

INTELLIGENCE NEEDED TO DEVISE THE BEST METHODS OF WORK: There is need of intelligence and educated ability to devise the best methods in farming, in building, and in every other department, that the worker may not labor in vain. (Fundamentals of Christian Education p. 316)

GOD WANTS EDUCATED ABILTIY TO BE APPLIED TO THE HUMBLEST KINDS OF WORK: There is science in the humblest kind of work, and if all would thus regard it, they would see nobility in labor. Heart and soul are to be put into work of any kind; then there is cheerfulness and efficiency. In agricultural or mechanical occupations men may give evidence to God that they appreciate His gift in the physical powers, and the mental faculties as well. Let the educated ability be employed in devising improved methods of work. This is what the Lord wants. There is honor in any class of work that is essential to be done. Let the law of God be made the standard of action, and it ennobles and sanctifies all labor. Faithfulness in the discharge of every duty makes the work noble, and reveals a character that God can approve. (Fundamentals of Christian Education p. 315)

THERE IS A GREAT NEED FOR INTELLIGENT MEN WHO WILL CULTIVATE THE SOIL: Many kinds of labor adapted to different persons may be devised. But the working of the land will be a special blessing to the worker. There is a great want of intelligent men to till the soil, who will be thorough. This knowledge will not be a hindrance to the education essential for business or for usefulness in any line. (Special Testimonies on Education p. 99)

EDUCATED ABILITY TO BE EMPLOYED IN DEVISING BETTER METHODS OF WORK: There is need of much more extensive knowledge in regard to the preparation of the soil. There is not sufficient breadth of view as to what can be realized from the earth. A narrow and unvarying routine is followed with discouraging results. Let the educated ability be employed in devising improved methods of work. This is just what the Lord wants. (Advocate March 1, 1901)

CULTIVATING LAND REQUIRES ALL THE TACT AND BRAINPOWER WE HAVE: The cultivation of our land requires the exercise of all the brainpower and tact we possess. The lands around us testify to the indolence of human beings. We hope to arouse to action the dormant senses. We hope to see intelligent farmers who will be rewarded for their earnest labor. The hand and heart must cooperate, bringing new and sensible plans into operation in the cultivation of the soil. (Testimonies to Ministers and Gospel Workers p. 242-244)

THOUGHT AND INTELLIGENCE NEEDED TO BE A FARMER: To develop the capacity of the soil requires thought and intelligence. Not only will it develop muscle, but capability for study, because the action of brain and muscle is equalized. (Special Testimonies on Education p. 99)

NECESSARY TO STUDY, PLAN AND EXPERIMENT TO SEE WHAT CAN BE DONE WITH THE SOIL: The young men should learn to cultivate the soil, and to raise whatever the land will produce. No one can tell what can be done with the soil until he has studied, planned, and experimented. (Pamphlet 151)

GOD WILL IMPRESS MINDS HOW TO EDUCATE THE PEOPLE IN AGRICULTURE AND MANUFACTURE HEALTH FOODS: The Lord will most surely impress minds in every place to devise means for the maintenance of the interests which are to feed the hungry, clothe the naked, and teach the ignorant, educating them in simple lines of book learning and in agriculture. He will give them wisdom to manufacture necessary, wholesome foods, which will be more needed in the Southern States than in any other part of America. He who feeds the ravens and cares for the wild beasts will give wisdom and skill, talent and ingenuity, for the production of wholesome foods, which are to be sold to the poor at as low a rate as possible. (Health Food Ministry p. 57)

INTELLIGENCE AND EDUCATED ABILITY NEEDED TO DEVISE THE BEST METHODS OF FARMING: Thou shalt love the Lord thy God with all thy heart, and with all thy soul, and with all thy mind, and with all thy strength (Mark 12:30). God desires the love that is expressed in heart service, in soul service, in the service of the physical powers. We are not to be dwarfed in any kind of service for God. Whatever He has lent us is to be used intelligently for Him.... There is need of intelligence and educated ability to devise the best methods in farming, in building, and in every other department, that the worker may not labor in vain.... It is the duty of every worker not merely to give his strength but his mind and intellect to that which he undertakes to do.... You can choose to become stereotyped in a wrong course of action because you have not the determination to take yourselves in hand and reform, or you may cultivate your powers to do the very best kind of service, and then you will find yourselves in demand anywhere and everywhere. You will be appreciated for all that you are worth. Whatsoever thy hand findeth to do, do it with thy might (Ecclesiastes 9:10). Not slothful in business; fervent in spirit; serving the Lord (Romans 12:11). (That I May Know Him p. 333)

— (6): Leadership Ability

ACTIVE PEOPLE (INCLUDING FARMERS) ARE WELL-SUITED TO SERVE IN CHURCH LEADERSHIP ROLES: Those in the church who have sufficient talent to engage in any of the various vocations of life, such as teaching, building, manufacturing, and farming, will generally be prepared to labor for the upbuilding of the church by serving on committees or as teachers in Sabbath-schools, engaging in missionary labor or filling the different offices connected with the church. (Review and Herald, January 1, 1880)

NOBLE, INTELLIGENT FARMERS TO BE INFLUENTIAL: There will be a new presentation of men as breadwinners, possessing educated, trained ability to work the soil to advantage. Their minds will not be overtaxed and strained to the uttermost with the study of the sciences. Such men will break down the foolish sentiments that have prevailed in regard to manual labor. An influence will go forth, not in loud-voiced oratory, but in real inculcation of ideas. We shall see farmers who are not coarse and rough and slack, careless of their apparel and of the appearance of their homes; but they will bring taste into farmhouses. Rooms will be sunny and inviting. We shall not see blackened ceilings, covered with cloth full of dust and dirt. Science, genius, intelligence, will be manifest in the home. The cultivation of the soil will be regarded as elevating and ennobling. Pure, practical religion will be manifested in treating the earth as God's treasure-house. The more intelligent a man becomes, the more should religious influence be radiating from him. And the Lord would have us treat the earth as a precious treasure, lent us in trust. (Special Testimonies for Ministers and Workers p. 19)

THOSE WHO ARE THE MOST WILLING TO TOIL AND SHOW INDUSTRY IN BUSINESS LINES WILL BE CHOSEN BY GOD TO DO HIS SERVICE WHEREVER THEIR LOT MAY BE CAST: A man's success in the ministry does not rest upon his excluding himself from useful labor, nor upon his popularity or indolence, but upon his willingness to labor in any position that seems to be duty. Those who are the most willing to toil and show industry in business lines, and who, themselves, plan and devise to be a help to others in branches of common toil, are the men who will be chosen by God to do Him service wherever their lot may be cast. They may be called upon with the help of others to build their own homes, or to build a church, or to do this alone, if they have a knowledge of how to handle tools. (Manuscript Release 19 p. 26)

THOSE WHO DON'T APPRECIATE THE VALUE OF AGRICULTURAL WORK SHOULD NOT PLAN FOR OUR SCHOOLS: Some do not appreciate the value of agricultural work. These should not plan for our schools, for they will hold everything from advancing in right lines. In the past their influence has been a hindrance. (Testimonies to the Church Vol. 6 p. 178)

THOSE WHO ARE NOT ADAPTED TO AGRICULTURAL WORK SHOULD NOT HAVE PLANNING ROLES IN OUR CONFERENCES, FOR THEY WILL HOLD BACK THE WORK: There are those who are not adapted to agricultural work. These should not devise and plan in our conferences, for they will hold everything from advancing in these lines. This has held our people from advancing in the past. If the land is cultivated, it will with the blessing of God, supply our necessities. (Letter 75, 1898 to E. A. Sutherland)

— (7): Perseverance, Discipline, and Patience

LESSONS OF PERSEVERANCE AND PATIENCE CONSTANTLY LEARNED FROM TILLING THE SOIL: From the tilling of the soil, lessons may constantly be learned. No one settles upon a raw piece of land with the expectation that it will at once yield a harvest. Diligent, persevering labor must be put forth in the preparation of the soil, the sowing of the seed, and the culture of the crop. So it must be in the spiritual sowing. The garden of the heart must be cultivated. The soil must be broken up by repentance. The evil growths that choke the good grain must be uprooted. As soil once overgrown with thorns can be reclaimed only by diligent labor, so the evil tendencies of the heart can be overcome only by earnest effort in the name and strength of Christ. (Education p. 111)

THE EARTH HAS TREASURES FOR THOSE WITH THE PERSEVERANCE, COURAGE, AND WILL TO GATHER: In the beginning, God looked upon all He had made, and pronounced it very good. The curse was brought upon the earth in consequence of sin. But shall this curse be multiplied by increasing sin? Ignorance is doing its baleful work. Slothful servants are increasing the evil by their lazy habits.... But the earth has blessings hidden in its depths for those who have courage and will and perseverance to gather its treasures. (Fundamentals of Christian Education pp. 315-327)

EVEN THE OXEN USED TO CLEAR LAND ARE PATIENT AND DISCIPLINED: Large tracts of beautiful land lie uncleared, unworked. The timber business has brought the settlers a meager pittance, and almost every day we see a drove of bullocks used to draw one, or sometimes two or three large logs. We count six, seven, or eight span, moving slowly along with their burden. Six span of bullocks were used to plow our land for cultivation. They are under discipline, and will move at a word and a crack of a whip, which makes a sharp report, but does not touch them. They wheel into line when it seems that they must get tangled up, but the creatures understand their business, and they plod patiently with the immense plow used to break up the unworked soil. (Manuscript Release p. 893)

ATTENTION TO THE LAWS INVOLVED NEEDED FOR SUCCESS: No one can succeed in agriculture or gardening without attention to the laws involved. The special needs of every variety of plant must be studied. Different varieties require different soil and cultivation, and compliance with the laws governing each is the condition of success. (Education p. 111-112)

ENCOURAGEMENT FOR BETTER, MORE DILIGENT FARMING: The talents entrusted to the keeping of those in the school have not been diligently put out to the exchangers. The character of much of the work has left an unfavorable impression upon the minds of unbelievers. It is time now to take up the work in faith and prayer with all the capabilities God has given. Cultivate the land and it will produce its treasures. Turn to God in faith, working as under the eye of the great Searcher of hearts. Let each worker encourage the one next to him, each holding up the hands of each, all yielding obedience to God's requirements. (A Place Called Oakwood p. 120)

— (8): Promptness and Responsibility

CULTIVATE PROMPTNESS AND RESPONSBILITY—EVEN IN SMALLER OR DISAGREEABLE DUTIES: There will be disagreeable tasks to be performed. Let no duty be overlooked, with the expectation that someone else will perform it. Let there be no superficial work done in any part of the school. Take hold of the forbidding task, and master it, and thus you will obtain a victory. The putting off even of little duties weakens the habits of promptness that should be encouraged. Cultivate the habit of seeing what ought to be done, and do it promptly. If a board is broken in the walk, do not leave it for someone else to repair. Let each one feel a responsibility for the care of the premises. Overcome natural indolence. Do not neglect the disagreeable things, supposing that they will be attended to by someone else. All these rules are important for the formation of right character. (A Place Called Oakwood p. 122)

TIMING, CAREFULNESS, AND COOPERATION WITH GOD NEEDED FOR BOTH FARMING AND SPIRITUAL MATTERS: God has given man land to be cultivated. But in order that the harvest may be reaped, there must be harmonious action between divine and human agencies. The plow and other implements of labor must be used at the right time. The seed must be sown in its season. Man is not to fail of doing his part. If he is careless and negligent, his unfaithfulness testifies against him. The harvest is proportionate to the energy he has expended. So it is in spiritual things. We are to be laborers together with God. Man is to work out his own salvation with fear and trembling, for it is God that worketh in him, both to will and to do of his good pleasure. There is to be co-partnership, a divine relation, between the Son of God and the repentant sinner. We are made sons and daughters of God. As many as received him, to them gave he power to become the sons of God. Christ provides the mercy and grace so abundantly given to all who believe in him. He fulfils the terms upon which salvation rests. But we must act our part by accepting the blessing in faith. God works and man works. Resistance of temptation must come from man, who must draw his power from God. Thus he becomes a co-partner with Christ. (Review and Herald, May 28, 1908)

THOUGHTFUL, DAILY ATTENTION—RATHER THAN IMPULSIVE ACTIONS— NEEDED TO GARNER THE HARVEST: The soil will not produce its riches when worked by impulse. It needs thoughtful, daily attention. It must be plowed often and deep, with a view to keeping out the weeds that take nourishment from the good seed planted. Thus those who plow and sow prepare for the harvest. None need stand in the field amid the sad wreck of their hopes. (Christ's Object Lessons p. 88)

LAXNESS NOT TO BE FOUND ATTACHED TO OUR SCHOOLS OR FARM: No laxness is to be allowed. The man who takes charge of the Huntsville School should know how to govern himself and how to govern others. The Bible teacher should be a man who can teach the students how to present the truths of the Word of God in public, and how to do house-to-house work. The business affairs of the farm are to be wisely and carefully managed. (Manuscript Release 2 p. 69)

— (9): Self-Sufficiency

Chapter #4: Blessings to be Obtained through Involvement in Farming

AGRICULTURAL WORK OPENS THE WAY TO BE SELF-SUPPORTING: Students should be given a practical education in agriculture. This will be of inestimable value to many in their future work. The training to be obtained in felling trees and in tilling the soil, as well as in literary lines, is the education that our youth should seek to obtain. Agriculture will open resources for self-support. Other lines of work, adapted to different students, may also be carried on. But the cultivation of the land will bring a special blessing to the workers. We should so train the youth that they will love to engage in the cultivation of the soil. (Counsels to Parents, Teachers, and Students p. 311)

BY RAISING THEIR OWN FOOD, THE POOR CAN BE COMPARATIVELY INDEPENDENT OF HARD TIMES: Much might be done in this country [Australia] if there were those who would settle in different localities and cultivate the land as they do in America. Then they would be comparatively independent of the hard times. I think this will be brought about. Most diligent search has been made for a tract of land of several hundred acres on which to locate the school, so that the students may have an opportunity to till the soil, and poor families may have a little piece of land on which to grow vegetables and fruit. This would go far toward sustaining them, and they would have a chance to school their children. But money matters are very close. The people are all hard pressed for means, and know not just what to do unless times change. We must live and have means to carry forward the work. (Review and Herald, May 29, 1894)

THOUSANDS WHO ARE IN THE CITIES SEEKING TO EARN A TRIFLE SHOULD BE EARNING A LIVING ON FARMS INSTEAD: He who taught Adam and Eve in Eden how to tend the garden would instruct men today. There is wisdom for him who holds the plow and plants and sows the seed. The earth has its concealed treasures, and the Lord would have thousands and tens of thousands working upon the soil who are crowded into the cities to watch for a chance to earn a trifle.... Those who will take their families into the country place them where they have fewer temptations. The children who are with parents that love and fear God are in every way much better situated to learn of the Great Teacher, who is the source and fountain of wisdom. They have a much more favorable opportunity to gain a fitness for the kingdom of heaven. (Adventist Home p. 143)

GOD'S PLAN FOR FREED SLAVES INCLUDED BIBLE AND AGRICULTURAL TRAINING SO THEY COULD EARN A LIVELIHOOD: Why should not Seventh-day Adventists become true laborers together with God in seeking to save the souls of the colored race? Instead of a few, why should not many go forth to labor in this long-neglected field? Where are the families who will become missionaries and who will engage in labor in this field? Where are the men who have means and experience so that they can go forth to these people and work for them just where they are? There are men who can educate them in agricultural lines, who can teach the colored people to sow seed and plant orchards. There are others who can teach them to read, and can give them an object lesson from their own life and example. Show them what you yourself can do to gain a livelihood, and it will be an education to them. Are we not called upon to do this very work? Are there not many who need to learn to love God supremely and their fellow men as themselves? In the Southern field are many thousands of people who have souls to save or to lose. Are there not many among those who claim to believe the truth who will go forth into this field to do the work for which Christ gave up His ease, His riches, and His life? (Review and Herald, November 26, 1895)

IMPORTANCE OF FAMILIES RAISING THEIR OWN PROVISIONS: Again and again the Lord has instructed that our people are to take their families away from the cities, into the country, where they can raise their own provisions; for in the future the problem of buying and selling will be a very serious one. We should now begin to heed the instruction given us over and over again: Get out of the cities into rural districts, where the houses are not crowded closely together, and where you will be free from the interference of enemies. (Manuscript Release 19 p. 229)

LET EACH CHILD HAVE A PLOT OF HIS OR HER OWN TO GARDEN. LESSONS LEARNED THERE WILL PAY GREATLY: If possible, let your home be out of the city, that your children may have ground to cultivate. Let them each have a piece of ground as their own, and as you teach them how to make a garden, how to prepare the soil for the seed, and the importance of keeping all the weeds pulled out, teach them how important it is to keep unsightly, injurious practices out of the life. Teach them to keep down wrong habits as they keep down the weeds in their gardens. It will take time to teach these lessons, but it will pay, yes, greatly pay. (Manuscript Release 10 p. 326)

MORE SCHOOLS TO BE ESTABLISHED. STUDENTS TO BE TAUGHT TO CULTIVATE THE SOIL AND MAKE THEMSELVES SELF-SUPPORTING. We must provide greater facilities for the education and training of the youth, both white and colored. We are to establish schools away from the cities, where the youth can learn to cultivate the soil, and thus help to make themselves and the school self-supporting. Let means be gathered for the establishment of such schools. In connection with these schools, work is to be done in mechanical and agricultural lines. All the different lines of work that the situation of the place will warrant are to be brought in. (Manuscript Release p. 264)

SELF-SUFFICIENCY NEEDED: The people should learn as far as possible to depend upon the products that they can obtain from the soil. In every phase of this kind of labor they can be educating the mind to work for the saving of souls for whom Christ has died. Ye are God's husbandry; ye are God's building. (Special Testimonies on Education p. 102)

THE ABILITY TO BE SELF-SUPPORTING CRUCIAL AND ALSO A BLESSING TO OTHERS: We are endeavoring to bring the colored people to that place where they can be self-supporting. The time will come when you will be able to escape many of the evils that will come upon the world because you have obtained a correct knowledge of how to plant and to build, and how to carry on various enterprises. This is why we want this land occupied and cultivated, why we want buildings put up. The students are to learn how to plant and to build and to sow. As they learn to do this, they will see a work before them which they will be very glad to have a part in. Opportunities will present themselves by which they can make themselves a blessing to those around them. (A Place Called Oakwood p. 7)

THOUSANDS COULD BE HELPED TO ACHIEVE SELF-SUPPORT THROUGH TILLING THE SOIL: Our schools could aid effectively in the disposition of the unemployed masses. Thousands of helpless and starving beings, whose numbers are daily swelling the ranks of the criminal classes, might achieve self-support in a happy, healthy, independent life if they could be directed in skillful, diligent labor in the tilling of the soil. (Education p. 220)

WE SHOULD HELP THE POOR FIND HOMES ON THE LAND: We are to help the poor find homes upon the land. Hard work, simple fare, close economy, often hardship and privation, would be their lot. But what a blessing would be theirs in leaving the city, with its enticements to evil, its turmoil and crime, misery and foulness, for the country's quiet and peace and purity. (Ministry of Healing p. 190-191)

GOD'S PLAN FOR SELF-SUFFICIENCY: In God's plan for Israel every family had a home on the land, with sufficient ground for tilling. Thus were provided both the means and the incentive for a useful, industrious, and self-supporting life. And no devising of men has ever improved upon that plan. To the world's departure from it is owing, to a large degree, the poverty and wretchedness that exists today. (Ministry of Healing p. 183-184)

TRAIN SELF-SUPPORTING WORKERS: In many places self-supporting missionaries can work successfully! It was as a self-supporting missionary that the apostle Paul labored in spreading the knowledge of Christ throughout the world. While daily teaching the gospel in the great cities of Asia and Europe, he wrought at the trade of a craftsman to sustain himself and his companions. (Ministry of Healing p. 154)

FORMATION OF SMALL COMPANIES GOD'S PLAN: The formation of small companies as a basis of Christian effort has been presented to me by One who cannot err....If in one place there are only two or three who know the truth, let them form themselves into a band of workers. Let thee keep their bond of union unbroken, pressing together in love and unity, encouraging one another to advance, each gaining courage and strength from the assistance of the others. As they work and pray in Christ's name, their numbers will increase; for the Savior says, 'If two of you shall agree on earth as touching anything that they shall ask, it shall be done for them of My Father which is in heaven.' Matthew 18:19 (Testimonies to the Church Vol. 7 p. 21- 22)

SELF-SUPPORTING MISSIONARIES TO MOVE FORWARD: The Macedonian cry is coming from every quarter. Shall men go to the 'regular lines' to see whether they will be permitted to labor, or shall they go out and work as best they can, depending on their own abilities and on the help of the Lord, beginning in a humble way and creating an interest in the truth in places in which nothing has been done to give the warning message? (Medical Ministry p. 321)

WORK WITH YOUR HANDS AND PROCLAIM THE MESSAGE OF WARNING: Young men, go forth into the places to which you are directed by the Spirit of the Lord. Work with your hands, that you may be self-supporting, and as you have opportunity proclaim the message of warning. (Medical Ministry p. 322)

AGRICULTURAL TRAINING AND SELF-SUPPORTING WORK: Students should be given a practical education in agriculture. Agriculture will open resources for self-support. Other lines of work, adapted to different students, may also be carried on. But the cultivation of the land will bring a special blessing to the workers. We should so train the youth that they will love to engage in the cultivation of the soil. (Counsels to Parents, Teachers and Students p. 311)

— (10): Speed and Efficiency

SLOW HABITS TO BE OVERCOME: The slow habits must be overcome. The man who is slow, and does his work at a disadvantage, is an unprofitable workman. His slowness is a defect that needs to be seen and corrected. He needs to exercise his intellect in planning how to use his time so as to secure the best results. When one is forever at work, and the work is never done, it is because mind and heart are not put into the work. It takes some persons ten hours to do that which another accomplishes readily in five. Such workmen do not bring tact and method into their labor. There is something to be learned every day as to how to improve in the manner of labor so as to get through the work, and have time for something else. (Fundamentals of Christian Education p. 316)

MIND AND HEART SPEED THE WORK: When one is forever at work, and the work is never done, it is because mind and heart are not put into the work (Fundamentals of Christian Education p. 316)

TACT AND METHOD NEEDED TO SPEED UP LABOR: It takes some persons ten hours to do that which another accomplishes readily in five. Such workmen do not bring tact and method into their labor. There is something to be learned every day as to how to improve in the manner of labor so as to get through the work, and have time for something else. (Fundamentals of Christian Education p. 316)

DEFTNESS, DISPATCH, AND ORDER: It is the duty of every Christian to acquire habits of order, thoroughness, and dispatch. There is no excuse for slow bungling at work of any character. When one is always at work and the work is never done, it is because mind and heart are not put into the labor. The one who is slow and who works at a disadvantage should realize that these are faults to be corrected. He needs to exercise his mind in planning how to use the time so as to secure the best results All who will, may overcome these fussy, lingering habits. In their work let them have a definite aim. Decide how long a time is required for a given task, and then bend every effort toward accomplishing the work in the given time. The exercise of the willpower will make the hands move deftly. (Christ's Object Lessons p. 344)

— (11): Thoroughness and Diligence

PLOW DEEPLY: The plow was only put in the depth of a few inches, the ground was not prepared for the seed, and the harvest was meager, corresponding to the superficial preparation that was given to the land. (Fundamentals of Christian Education p. 368)

DILIGENCE IN NATURE: Everything in nature is diligent, and moving steadily onward, setting us an example. Notwithstanding the plants and shrubs are thirsting for showers, yet they cannot stop to complain, and cease their efforts to flourish. They obey nature's laws, to do the very best they can under every circumstance. They thirst to be refreshed with water, yet they strike their roots down deeper, reaching down far to gather the moisture, that they may retain life, freshness, and beauty. (The Health Reformer, June, 1871)

HONOR GOD IN ALL THINGS: Aim to honor God in everything, always and everywhere. Carry your religion into everything. Be thorough in whatever you undertake. (Testimonies to the Church Vol. 2 p. 262)

THOUGHTFUL, DAILY ATTENTION: The soil will not produce its riches when worked by impulse. It needs thoughtful, daily attention. (Christ's Object Lessons p. 88)

THE NEED FOR CULTIVATION: A field left uncultivated speedily produces a rank growth of thistles and tangled vines, which exhaust the soil and are worthless to the owner. The ground is full of seeds blown and carried by the wind from every quarter, and if it is left uncultivated, they spring up to life spontaneously, choking every precious fruit-bearing plant that is struggling for existence. (My Life Today p. 83)

PRIDE IN KEEPING OUT WEEDS: You may have a pride in keeping out every weed, and may watch with interest the beautiful development of every leaf and opening bud and flower, and be charmed with the miracles of God in nature. (Health Reformer, June, 1871)

PAINSTAKING EFFORT NEEDED: Oranges, lemons, prunes, peaches, and many other varieties can be obtained; for the Lord's world is productive, if painstaking effort is put forth. (Counsels on Diet and Foods p. 311)

METHOD AND TACT REQUIRED: This real, earnest work calls for strength of intellect as well as of muscle. Method and tact are required even to raise fruits and vegetables successfully. (Fundamentals of Christian Education p. 322)

ALL ENERGY NEEDED: Exert your God-given abilities, and bring all your energies into the development of the Lord's farm. (Testimonies to the Church Vol. 6 p. 192)

Blessing #2: Unseen Advantages and Heaven-planned Partnerships

UNSEEN ADVANTAGES AWAIT IN INDUSTRIAL TRAINING: We must not be narrow in our plans. In industrial training there are unseen advantages, which cannot be measured or estimated. Let no one begrudge the effort necessary to carry forward successfully the plan that for years has been urged upon us as of primary importance. (Pamphlet 164 p. 37)

LABORING TOGETHER WITH GOD: When students employ their time and strength in agricultural work, in heaven it is said of them, 'Ye are laborers together with God.' (1 Corinthians 3:9) (Testimonies to the Church Vol. 6 p. 187)

ANGELS WORKING WITH YOU: Angels of God are right around you. They will minister to the very earth, causing it to give forth its treasure. (The Upward Look p. 130)

GOD'S IMMEDIATE AGENCY: The same God who guides the planets works in the fruit orchard and in the vegetable garden. It is through God's immediate agency that every bud bursts into blossom. (Testimonies to the Church Vol. 6 p. 186)

A CLOSE CONNECTION WITH HEAVEN: "The things of earth are more closely connected with heaven, and are more directly under the supervision of Christ, than many realize…Human and divine agencies are to combine in temporal as well as spiritual achievements. They are to be united in all human pursuits, in mechanical and agricultural labors, in mercantile and scientific enterprises. (Counsels to Parents, Teachers and Students p. 277)

HEAVENLY ANGELS WATCHING THAT FARM: I pray that all who shall visit the school farm may see, by the united efforts of students and teachers, that the best kind of education is being given. Heavenly angels are watching that farm, desiring that it may be so worked by the students, that the students themselves, under the direction of wise teachers, shall show that improvement in their characters which God desires to see. (Special Testimonies, Series B, number 12, p. 16)

Chapter #4: Blessings to be Obtained through Involvement in Farming

SEEK DIVINE WISDOM: How can he get wisdom that holdeth the plow, and driveth oxen?—by seeking her as silver, and searching for her as for hid treasures. 'For his God doth instruct him to discretion, and doth teach him.' (Fundamentals of Christian Education p. 326)

GOD WILL INSTRUCT US TODAY: He who taught Adam and Eve in Eden how to tend the garden, would instruct men today. There is wisdom for him who holds the plow, and plants and sows the seed. (Fundamentals of Christian Education p. 326)

Blessing #3: Physical Benefits to Students and Others

CHILDREN MUCH IN THE OPEN AIR TO HAVE VIVACITY, CHEERFULNESS, AND HEALTH: In order for children and youth to have health, cheerfulness, vivacity, and well-developed muscles and brains, they should be much in the open air, and have well-regulated employment and amusement. Children and youth who are kept at school and confined to books cannot have sound physical constitutions. The exercise of the brain in study, without corresponding physical exercise, has a tendency to attract the blood to the brain, and the circulation of the blood through the system becomes unbalanced. The brain has too much blood, and the extremities too little. There should be rules regulating the studies of children and youth to certain hours, and then a portion of their time should be spent in physical labor. And if their habits of eating, dressing, and sleeping are in accordance with physical law, they can obtain an education without sacrificing physical and mental health. (Counsels for the Church p. 209)

OCCUPATIONS REQUIRING EXERCISE IN THE OPEN AIR ARE THE BEST FOR CLEAR THINKING AND HEALTH: The various trades and occupations have to be learned, and they call into exercise a great variety of mental and physical capabilities; the occupations requiring sedentary habits are the most dangerous, for they take men away from the open air and sunshine, and train one set of faculties, while other organs are becoming weak from inaction. Men carry on their work, perfect their business, and soon lie down in the grave. Much more favorable is the condition of one whose occupation keeps him in the open air, exercising his muscles, while the brain is equally taxed, and all the organs have the privilege of doing their work. To those who can live outside of the cities, and labor in the open air, beholding the works of the great Master Artist, new scenes are continually unfolding. As they make the book of nature their study, a softening, subduing influence comes over them; for they realize that God's care is over all, from the glorious sun in the heavens to the little brown sparrow or the tiniest insect that has life. The Majesty of heaven has pointed us to these things of God's creation as an evidence of his love. He who fashioned the flowers has said: Behold the lilies of the field, how they grow; they toil not, neither do they spin; and yet I say unto you that even Solomon in all his glory was not arrayed like one of these. Wherefore if God so clothe the grass of the field, which today is, and tomorrow is cast into the oven, shall he not much more clothe you, O ye of little faith? The Lord is our teacher, and under his instruction we may learn the most precious lessons from nature. (Special Testimonies on Education p. 95)

THE BODY DESIGNED FOR ACTION: The whole body is designed for action; and unless the physical powers are kept in health by active exercise, the mental powers cannot long be used to their highest capacity. (Education p. 207)

HEALTH-GIVING, LIFE-GIVING IMPACT OF NATURE: Nature is God's physician. The pure air, the glad sunshine, the flowers, the trees, the orchards and vineyards, and outdoor exercise amid these surroundings, are health-giving, life-giving. (Testimonies for the Church Vol. 7 p. 79)

INVIGORATING THE SYSTEM: The pure air has in it health and life. As it is breathed in, it has an invigorating effect on the whole system. (Medical Ministry p. 232)

LIFE-GIVING PROPERTIES IN THE TREES: There are life-giving properties in the balsam of the pine, in the fragrance of the cedar and the fir. And there are other trees that are health-promoting. Let no such trees be ruthlessly cut down. Cherish them where they are abundant, and plant more where there are but few. (Testimonies to the Church Vol. 7 p. 77)

FLOWERS AND ORCHARDS AWAKEN HOPE AND JOY: Under these influences, combined with the influence of careful treatment and wholesome food, the sick find health. The feeble step recovers its elasticity. The eye regains its brightness. The hopeless become hopeful. The once despondent countenance wears an expression of cheerfulness.

Chapter #4: Blessings to be Obtained through Involvement in Farming

The complaining tones of the voice give place to tones of content. (Testimonies to the Church Vol. 7 p. 86)

HEALTH-GIVING PROPERTIES OF FLOWERS AND TREES: The Lord is leading my mind to the health-giving properties of the flowers and trees. (Adventist Home p. 147)

HEALTH-GIVING QUALITIES OF EXERCISE IN ORCHARD AND VINEYARD: The pure air, the glad sunshine, the beautiful flowers and trees, the orchards and vineyards, and outdoor exercise amid these surroundings, are health-giving—the elixir of life. (Testimonies to the Church Vol. 7 pp. 76-77)

BECAUSE OF THE EXERCISE AND FRESH AIR THEY OBTAIN, FARMERS ARE GENERALLY HEALTHY: Where useful labor is combined with study, there is no need of gymnastic exercises; and much more benefit is derived from work performed in the open air than from indoor exercise. The farmer and the mechanic each have physical exercise; yet the farmer is much the healthier of the two, for nothing short of the invigorating air and sunshine will fully meet the wants of the system. The farmer finds in his labor all the movements that were ever practiced in the gymnasium. And his movement room is the open fields; the canopy of heaven is its roof, and the solid earth its floor. A farmer who is temperate in all his habits usually enjoys good health. His work is pleasant; and his vigorous exercise causes full, deep, and strong inspirations and exhalations, which expand the lungs and purify the blood, sending the warm current of life bounding through arteries and veins. (Signs of the Times, August 26, 1886)

COUNTRY LIVING DEVELOPS HEALTH OF MIND AND BODY, TENDING PLANTS REFINDS AND ENNOBLES: To live in the country would be very beneficial to them [children]; an active, out-of-door life would develop health of both mind and body. They should have a garden to cultivate, where they might find both amusement and useful employment. The training of plants and flowers tends to the improvement of taste and judgment, while an acquaintance with God's useful and beautiful creations has a refining and ennobling influence upon the mind, referring it to the Maker and Master of all. (Testimonies to the Church Vol. 4 p. 136)

WORKING OUTDOORS BRINGS HEALTH AND HEALING: Let houses be built for families who have not a firm hold of life. Let men and women work in fields and orchard and garden. This will bring health and strength to nerve and muscle. Living indoors and cherishing invalidism is a very poor business. If those who are sick will give nerves and muscles and sinews proper exercise in the open air, their health will be renewed. (Manuscript Release 19 p. 230)

TENDING A SMALL PLOT IMPROVES PHYSICAL HEALTH OF INVALIDS: Live, dear invalid friends, while you do live, and train yourselves to shed fragrance like the fresh flowers. If you are burdened and weary, you need not curl up like leaves upon a withered branch. Cheerfulness and a clear conscience are better than drugs, and will be an effective agent in your restoration to health. In order for you to be cheerful, you should have exercise. You should have something useful to do. Invalid sisters should have something to call them out of doors, to work in the ground. This was the employment given by God to our first parents. God knew that employment was necessary to happiness. You should have a spot of ground to claim as yours, to tend and cultivate. You may have a pride in keeping out every weed, and may watch with interest the beautiful development of every leaf and opening bud and flower, and be charmed with the miracles of God seen in nature. As you view the shrubs and flowers, remember God loves the beautiful in nature. As you watch the harmonious colors of the various beautiful-tinted flowers of June, bear in mind that God loves the beautiful in human nature formed in his image. A pure, harmonious character, a sunny temper, reflecting light and cheerfulness, glorifies God, and benefits humanity. Inspiration tells us that a meek and quiet spirit in the sight of God is of great price. (Health Reformer, June 1, 1871)

BENEFITS OF WORK TO MUSCLE AND BRAIN: Work of brain and muscle is beneficial. Each faculty of the mind and each muscle of the body has its distinctive office, and all require exercise to develop them and give them healthful vigor. Each wheel in the living mechanism must be brought into use. The whole organism needs to be constantly exercised in order to be efficient and meet the object of its creation. (Christ Triumphant p. 20)

HEALTH BENEFITS FROM THE TREES: In a certain place, preparations were being made to clear the land for the erection of a sanitarium. Light was given that there is health in the fragrance of the pine, the cedar, and the fir. And there are several other kinds of trees that have medicinal properties that are health-promoting. Let not such trees be ruthlessly cut down. Better change the site of the [sanitarium] building than cut down these evergreen trees…I ask you not to cut away your pine trees. They will be a blessing to many. Let them live. I want to say to you, my brethren and sisters, that you have my prayers and my sympathy in your work. Remember that you are trees in the garden of the Lord, and that the divine protection is round about you. The more visible the line of demarcation between the flowers of God and the briar thorn of Satan's planting, the more the Lord is glorified. (Spalding and Magan Collection p. 228-229)

REGULAR PHYSICAL EXERCISE—SUCH AS THOSE PRACTICED BY A FARMER—WILL GIVE STUDENTS A NEW HOLD ON LIFE: In what contrast to the habits of the active farmer are those of the student who neglects physical exercise. The student sits day after day in a close room, bending over his desk or table, his chest contracted, his lungs crowded. His brain is taxed to the utmost, while his body is inactive. He cannot take full, deep inspirations; his blood moves sluggishly; his feet are cold, his head hot. How can such a person have health? It is not hard study that is destroying the health of students, so much as it is their disregard of nature's laws. Let them take regular exercise that will cause them to breathe deep and full, and they will soon feel that they have a new hold on life. (Signs of the Times, August 26, 1886)

OUTDOOR WORK A GREAT BENEFIT TO THE SICK: Many act as if health and disease were things entirely independent of their conduct, and entirely outside their control. They do not reason from cause to effect, and submit to feebleness and disease as a necessity. Violent attacks of sickness they believe to be special dispensations of Providence, or the result of some overruling, mastering power; and they resort to drugs as a cure for the evil. But the drugs taken to cure the disease weaken the system. If those who are sick would exercise their muscles daily, women as well as men, in outdoor work, using brain, bone, and muscle proportionately, weakness and languor would disappear. Health would take the place of disease, and strength the place of feebleness. (Manuscript Release 19 p. 230)

FARMING A CURE FOR DISEASE—IF DONE IN A BALANCED WAY: There is a large field for you to work in. Both of you can give short lectures in the parlor at stated times, which will be select but plain, upon the human body and how to treat this wonderful house the Lord has given us, which will aid you in your work as physicians as nothing else can. The people are ignorant, and need to be enlightened on almost every point of how to treat their own bodies. Then there will not need to be a dwelling upon the delicate diseases nearly as much. Tell those who are sick that if the hosts of those who are dyspeptics and consumptives could turn farmers they might overcome disease, dispense with drugs and doctors, and recover health. But farmers themselves must get educated to give heed to the laws of life and health by regulating their labor, even if there is some loss in their grain or the harvesting of crops. Farmers work too hard and too constantly, and violate the laws of God in their physical nature. This is the worst kind of economy. For a day he may accomplish more, yet in the end he is a loser by his ill management of himself. . . . (Manuscript Release 13 p. 371)

HEALTH AND HAPPINESS TO BE FOUND IN THE STUDY OF NATURE: A return to simpler methods will be appreciated by the children and youth. Work in the garden and field will be an agreeable change from the wearisome routine of abstract lessons, to which their young minds should never be confined. To the nervous child, who finds lessons from books exhausting and hard to remember, it will be especially valuable. There is health and happiness for him in the study of nature; and the impressions made will not fade out of his mind, for they will be associated with objects that are continually before his eyes. (Testimonies to the Church Vol. 6 p. 179)

MANY WHO REGARD PHYSICAL WORK AS DEGRADING LOSE THEIR HEALTH THROUGH LACK OF PHYSICAL EXERCISE: Men who have good physical powers should educate themselves to think as well as to act, and not depend upon others to be brains for them. It is a popular error with a large class, to regard work as degrading. Therefore young men are very anxious to educate themselves to become teachers, clerks, merchants, lawyers, and to occupy almost any position that does not require physical labor. Young women regard housework as demeaning. And although the physical exercise required to perform household labor, if not too severe, is calculated to promote health, they will seek for education that will fit them to become teachers, clerks, or learn some trade which confines them in-doors to sedentary employment. The bloom of health fades from their cheeks, and disease fastens upon them, because they are robbed of physical exercise; and their habits are perverted generally, because it is fashionable. They enjoy delicate life, which is feebleness and decay. (Christian Education p. 21)

ELEVATING AND ENNOBLING WORK: The cultivation of the soil will be regarded as elevating and ennobling. Pure, practical religion will be manifested in treating the earth as God's treasure house. (Testimonies to Ministers and Gospel Workers p. 245)

Chapter #4: Blessings to be Obtained through Involvement in Farming

INACTION IN HUMAN MACHINERY BRINGS SUFFERING AND DISEASE: Let those who are sick do all in their power, by correct practice in eating, drinking, and dressing, and by taking judicious exercise, to secure recovery of health. Let the patients who come to our sanitariums be taught to cooperate with God in seeking health. Ye are God's husbandry, ye are God's building. God made nerve and muscle in order that they might be used. It is the inaction of the human machinery that brings suffering and disease. (Manuscript Release 19 p. 231)

WHEN WE FOLLOW A COURSE THAT BRINGS WEAKNESS AND DISEASE, WE DISHONOR GOD: In this world we are to obtain a fitness for the higher world. God has left a trust with us, and he expects us to use all our faculties in helping and blessing our fellowmen. He calls for our best affections, our highest powers, and he is dishonored when we follow a course that brings weakness and disease upon the physical and mental powers. (Manuscript Release 3 114)

PHYSICAL LABOR CONTRIBUTES TO HAPPINESS: God ordained that the beings He created should work. Upon this their happiness depends. No one in the Lord's great domain of creation was made to be a drone. Our happiness increases and our powers develop as we engage in useful employment. (Life Sketches p. 87)

A REGULAR SCHEDULE FOR PHYSICAL LABOR NEEDED: The health cannot be preserved unless some portion of each day is given to muscular exertion in the open air. Stated hours should be devoted to manual labor of some kind. (Fundamentals of Christian Education p. 146)

EDUCATION ENHANCED BY PHYSICAL LABOR: Labor is a blessing. It is impossible for us to enjoy health without labor. All the faculties should be called into use that they may be properly developed and that men and women may have well-balanced minds. If the young had been given a thorough education in the different branches of labor, if they had been taught labor as well as the sciences, their education would have been of greater advantage to them. (Testimonies to the Church Vol. 3 p. 155)

Blessing #4: Spiritual Benefits to Students and Others

SPECIAL BLESSINGS CONFERRED ON STUDENTS WHO TILL THE SOIL: Other lines of work, adapted to different students, may also be carried on. But the cultivation of the land will bring a special blessing to the workers. We should so train the youth that they will love to engage in the cultivation of the soil. (Counsels to Parents, Teachers and Students p. 311)

THE RELATIONSHIP OF PHYSICAL LABOR TO REDEMPTION: At the creation, labor was appointed as a blessing. It meant development, power, happiness. The changed condition of the earth through the curse of sin has brought a change in the conditions of labor; yet though now attended with anxiety, weariness, and pain, it is still a source of happiness and development. And it is a safeguard against temptation. Its discipline places a check on self-indulgence, and promotes industry, purity, and firmness. Thus it becomes a part of God's great plan for our recovery from the fall. (Education p. 214)

STUDENTS WHO WORK IN A GARDEN ARE WORKING TOGETHER WITH GOD: The God of nature is perpetually at work. His infinite power works unseen, but manifestations appear in the effects which the work produces. The same God who guides the planets works in the fruit orchard and in the vegetable garden. He never made a thorn, a thistle, or a tare. These are Satan's work, the result of degeneration, introduced by him among the precious things; but it is through God's immediate agency that every bud bursts into blossom. When He was in the world in the form of humanity, Christ said: My Father worketh hitherto, and I work. John 5:17. So when the students employ their time and strength in agricultural work, in heaven it is said of them, Ye are laborers together with God. 1 Corinthians 3:9. (Australasian Union Conference Record, July 31, 1899)

UNDREAMED OF TREASURES OPENED UP THROUGH CULTIVATING THE SOIL: In the cultivation of the soil the thoughtful worker will find that treasures little dreamed of are opening up before him. No one can succeed in agriculture or gardening without attention to the laws involved. The special needs of every variety of plant must be studied. Different varieties require different soil and cultivation, and compliance with the laws governing each is the condition of success. The attention required in transplanting, that not even a root fiber shall be crowded or misplaced, the care of the young plants, the pruning and watering, the shielding from frost at night and sun by day, keeping out weeds, disease, and insect pests, the training and arranging, not only teach important lessons concerning the development of character, but the work itself is a means of development. In cultivating carefulness, patience, attention to detail, obedience to law, it imparts a most essential training. The constant contact with the mystery of life and the loveliness of nature, as well as the tenderness called forth in ministering to these beautiful objects of God's creation, tends to quicken the mind and refine and elevate the character. (Education p. 111)

USEFUL MANUAL LABOR—INCLUDING AGRICULTURE—TO CALM, UPLIFT, AND ENNOBLE STUDENTS: A constant strain upon the brain while the muscles are inactive, enfeebles the nerves and gives to students an almost uncontrollable desire for change and exciting amusements. When they are released, after being confined to study several hours each day, they are nearly wild. Many have never been controlled at home. They have been left to follow inclination, and they think that the restraint of the hours of study is a severe tax upon them; and because they have nothing to do after study hours, Satan suggests sport and mischief for a change. Their influence over other students is demoralizing. . . .Had there been agricultural and manufacturing establishments connected with our schools, and had competent teachers been employed to educate the youth in the different branches of study and labor, devoting a portion of each day to mental improvement and a portion to physical labor, there would now be a more elevated class of youth to come upon the stage of action, to have influence in molding society. Many of the youth graduated from such institutions would come forth with stability of character. They would have perseverance, fortitude, and courage to surmount obstacles, and such principles that they would not be swayed by a wrong influence however popular. (Counsels to Parents, Teachers and Students p. 288)

CONTEMPLATION OF NATURE DRAWS THE HEART AWAY FROM WORLDLY AMUSEMENTS: If the frivolous and pleasure-seeking will allow their minds to dwell upon the real and true, the heart cannot but be filled with reverence, and they will adore the God of nature. The contemplation and study of God's character as revealed in His created works will open a field of thought that will draw the mind away from low, debasing, enervating amusements. The knowledge of God's works and ways we can only begin to obtain in this world; the study will be continued throughout eternity. God has provided for man subjects of thought which will bring into activity every faculty of the mind. We may read the character of the Creator in the heavens above and the earth beneath, filling the heart with gratitude and thanksgiving. Every nerve and sense will respond to the expressions of God's love in His marvelous works. (Child Guidance p. 50)

TEACH THE YOUTH TO CONSIDER THE WORKS OF NATURE AND THEY WILL BE SPIRITUALLY BLESSED: Educate the children and youth to consider the works of the great Master Artist, and to imitate the attractive graces of nature in their character building. As the love of God wins their hearts, let them weave into their lives the beauty of holiness. So shall they use their capabilities to bless others and honor God. (Special Testimonies on Education p. 62)

BLESSINGS FOR YOUNG STUDENTS TENDING THE BEAUTIFUL THINGS OF NATURE: So also a new interest may be given to the work of the garden or the excursion in field and wood, as the pupils are encouraged to remember those shut in from these pleasant places and to share with them the beautiful things of nature. The watchful teacher will find many opportunities for directing pupils to acts of helpfulness. By little children especially the teacher is regarded with almost unbounded confidence and respect. Whatever he may suggest as to ways of helping in the home, faithfulness in the daily tasks, ministry to the sick or the poor, can hardly fail of bringing forth fruit. And thus again a double gain will be secured. The kindly suggestion will react upon its author. Gratitude and co-operation on the part of the parents will lighten the teacher's burden and brighten his path. (Counsels on Health p. 192)

THOSE WHO WORK WITH THE LAND WILL RECEIVE BLESSINGS AND EDUCATION IN THE THINGS OF GOD: The blessing of the Lord will rest upon those who thus work the land, learning spiritual lessons from nature. In cultivating the soil the worker knows little what treasures will open up before him. While he is not to despise the instruction he may gather from minds that have had an experience, and from the information that intelligent men may impart, he should gather lessons for himself.

This is a part of his training. The cultivation of the soil will prove an education to the soul. (Christ's Object Lessons p. 88)

WHILE THE HUMAN SOWER IS PLANTING SEEDS IN THE GROUND, THE DIVINE SOWER WILL BE PLANTING SEEDS IN HIS HEART: He who causes the seed to spring up, who tends it day and night, who gives it power to develop, is the Author of our being, the King of heaven, and He exercises still greater care and interest in behalf of His children. While the human sower is planting the seed to sustain our earthly life, the Divine Sower will plant in the soul the seed that will bring forth fruit unto life everlasting. (Christ's Object Lessons p. 89)

FARMING IS NOBLE, MANLY, AND BRINGS WITH IT SPIRITUAL BLESSINGS: The farmer and his sons have the open book of nature before them, and they should learn that farming is a noble occupation, when the work is done in a proper manner. The opinion that prevails that farming degrades the man, is erroneous. The earth is God's own creation, and he calls it very good. The hands may become hard and rough, but this hardness need not extend to the soul. The heart need not become careless, nor the soul defiled. The effeminate paleness may be tanned from the countenance, but the testimony of health is seen in the red and brown of the complexion. Christlikeness may be preserved in the farmer's life. Men may learn, in cultivating the soil, precious lessons about the cultivation of the Spirit. (Signs of the Times, August 13, 1896)

Blessing #5: Improved Thinking Ability and Vigor of Thought

FOR HEALTHIER BRAINS, ALL SHOULD BE TAUGHT TO WORK WITH THEIR HANDS: Men, women, and children should be educated to labor with their hands. Then the brain will not be overtaxed to the detriment of the whole organism. . . . (Manuscript Release 11 p. 174)

STUDENTS WHO ENGAGE IN DAILY, PHYSICAL LABOR WILL GAIN ELASTICITY OF SPIRIT AND VIGOR OF THOUGHT: We desire that our children should study to the best advantage. In order to do this, employment should be given them which would call into exercise the muscles. Daily, systematic labor should constitute a part of the education of youth even at this late period. Much can now be gained in connecting labor with schools. The students will acquire, in following this plan, elasticity of spirit and vigor of thought, and can accomplish more mental labor, in a given time, than they could by study alone. And they can leave their schools with their constitutions unimpaired, with strength and courage to persevere in any position in which the providence of God may place them. (Christian Education p. 22)

STUDY OF THE NATURAL WORLD STRENGTHENS THE MIND: The glory of God is displayed in His handiwork. Here are mysteries that the mind will become strong in searching out. Minds that have been amused and abused by reading fiction may in nature have an open book, and read truth in the works of God around them. All may find themes for study in the simple leaf of the forest tree, the spires of grass covering the earth with their green velvet carpet, the plants and flowers, the stately trees of the forest, the lofty mountains, the granite rocks, the restless ocean, the precious gems of light studding the heavens to make the night beautiful, the exhaustless riches of the sunlight, the solemn glories of the moon, the winter's cold, the summer's heat, the changing, recurring seasons, in perfect order and harmony, controlled by infinite power; here are subjects which call for deep thought, for the stretch of the imagination. (Child Guidance p. 49)

Blessing #6: Financial Success

— Financial Blessing #1: Bountiful Harvests through Diligent Farming

DILIGENT EFFORT IN FARMING WOULD PRODUCE GOOD RESULTS: If the people in this country would take the same pains in cultivating as in America, they would be able to grow as excellent fruit, grains, and vegetables as are raised there. If they would put forth the same effort, they might take the wild land in hand, and plough and sow it with grass seed for grazing cattle. (Manuscript Release 8 p. 135)

WITH DILIGENCE AND INTELLIGENCE FARMING CAN BE PROFITABLE: Fathers should train their sons to engage with them in their trades and employments.

Farmers should not think that agriculture is a business that is not elevated enough for their sons. Agriculture should be advanced by scientific knowledge. Farming has been pronounced unprofitable. People say that the soil does not pay for the labor expended upon it, and they bemoan the hard fate of those who till the soil. In this country (Australia) many have given up the idea that the land will pay for working it, and thousands of acres lie unimproved. But should persons of proper ability take hold of this line of employment, and make a study of the soil, and learn how to plant, to cultivate, and to gather in the harvest, more encouraging results might be seen. Many say, We have tried agriculture, and know what its results are, and yet these very ones need to know how to cultivate the soil, and to bring science into their work. Their plowshares should cut deeper, broader furrows, and they need to learn that in tilling the soil they need not become common and coarse in their natures. Let them learn to bring religion into their work. Let them learn to put in the seed in its season, to give attention to vegetation, and to follow the plan that God has devised. (Signs of the Times, August 13, 1896)

DILIGENT WORKERS WILL BE SUSTAINED BY GOD: There is need of intelligent and educated ability to devise the best methods in farming, in building, and in every other department, that the worker may not labor in vain. God, who has made the world for the benefit of man, will provide means from the earth to sustain the diligent worker. (Advocate, March 1, 1901)

DILIGENT LABOR AND WISE USE OF TIME TO LIFT PEOPLE FROM POVERTY AS THEY SUCCESSFULLY CULTIVATED THEIR LAND: God's entrusted talents are not to be hid under a bushel or under a bed. Ye are the light of the world, Christ said. Matthew 5:14. As you see families living in hovels, with scant furniture and clothing, without tools, without books or other marks of refinement about their homes, will you become interested in them, and endeavor to teach them how to put their energies to the very best use, that there may be improvement, and that their work may move forward? It is by diligent labor, by putting to the wisest use every capability, by learning to waste no time, that they will become successful in improving their premises and cultivating their land. (Testimonies to the Church Vol. 6 p. 188)

WITH DILIGENCE, FAITHFULNESS, AND THE BLESSING OF GOD— FARMERS CAN REAP A GOOD HARVEST: Because of the slack, slipshod way the landholders cultivate their farms, nothing flourishes as it should, and the impression made upon those who view the land is that it is too poor to yield a good crop. I have been anxious that the land should be taken in hand and thoroughly worked. Even the orange trees are left to grow up amid the grass, as wild trees grow. But where such immense trees flourish as flourish here, many of them growing up perfectly straight toward heaven, I am convinced that with the blessing of God, with diligence and faithfulness in working the land, farmers might produce gratifying results, and in return for the labor put forth they might reap a good harvest. (Manuscript Release 13 p. 349)

FINANCIAL HARDSHIP MAY BE THE LOT OF EVERY BELIEVER, BUT THE WAY CAN BE MUCH SMOOTHER THROUGH THE KNOWLEDGE OF PLANTING AND PRACTICAL TRADES: Privation may be the lot of every soul who now believes and obeys the truth. Christ has told us that we will have reproach. If persecution for the truth's sake is to come, it is important that every line of work become familiar to us, that we and our families may not suffer through lack of knowledge. We can and should have tact and knowledge in trades, in building, in planting, and in sowing. A knowledge of how to cultivate the land will make rough places much smoother. This knowledge will be counted a great blessing, even by our enemies. (Manuscript Release 19 p. 26)

THE HARVEST WILL REWARD THE PAINSTAKING EFFORTS MADE: Let the students call all their faculties of discernment to bear upon this subject. Let their skills interpret the figures used. The earth has to be worked to bring out its varied properties favorable to the growth of the seed and fruit. But the harvest will reward the painstaking efforts made in a supply of food for the necessities of man. . . . (Manuscript Release 860)

WHEN THE LAND IS CULTIVATED DILIGENTLY, CHEERFULLY, HOPEFULLY, GRATEFULLY—GOD WILL BLESS: If the land is cultivated, it will, with the blessing of God, supply our necessities. We are not to be discouraged about temporal things because of apparent failures, nor should we be disheartened by delay. We should work the soil cheerfully, hopefully, gratefully, believing that the earth holds in her bosom rich stores for the faithful worker to garner, stores richer than gold or silver. The niggardliness laid to her charge is false witness. With proper, intelligent cultivation the earth will yield its treasures for the benefit of man. The mountains and hills are changing; the earth is waxing old like a garment; but the blessing of God, which spreads a table for His people in the wilderness, will never cease. (Testimonies to the Church Vol. 6 p. 178)

GOD WILL BLESS THOSE WHO FOLLOW HIS PLAN: Many do not see the importance of having land to cultivate, and of raising fruit and vegetables, that their tables may be supplied with these things. I am instructed to say to every family and every church, God will bless you when you work out your own salvation with fear and trembling, fearing lest, by unwise treatment of the body, you will mar the Lord's plan for you. (Manuscript Release 19 p. 230)

INDEPENDENCE AND A FAIR RETURN FOR LABOR AVAILABLE THROUGH AGRICULTURE: Let the teacher call attention to what the Bible says about agriculture: that it was God's plan for man to till the earth; that the first man, the ruler of the whole world, was given a garden to cultivate; and that many of the world's greatest men, its real nobility, have been tillers of the soil. Show the opportunities in such a life. The wise man says, The king himself is served by the field. Ecclesiastes 5:9. Of him who cultivates the soil the Bible declares, His God doth instruct him to discretion, and doth teach him. Isaiah 28:26. And again, Whoso keepeth the fig tree shall eat the fruit thereof. Proverbs 27:18. He who earns his livelihood by agriculture escapes many temptations and enjoys unnumbered privileges and blessings denied to those whose work lies in the great cities. And in these days of mammoth trusts and business competition, there are few who enjoy so real an independence and so great certainty of fair return for their labor as does the tiller of the soil. (Education 219)

GOD WILL BLESS THE EFFORTS OF, AND GIVE WISDOM TO, FARM MANAGERS AND TEACHERS WHO WORK WITH THE HOLY SPIRIT: If the managers of this farm and the teachers in the school will receive the Holy Spirit to work with them, they will have wisdom in their management, and God will bless their labors. (Australasian Union Conference Record, *July 31, 1899).*

— Financial Blessing #2: Good Yields from Unpromising Ground

LAND THAT SEEMED USELESS WOULD YIELD THROUGH HARD WORK AND CULTIVATION: The people about here have raised no vegetables, and but little fruit, except a few oranges and lemons that are not cultivated, and I have seen a few peach trees. Land is profitless, but in the land boom it cost eight pounds an acre, some of which now sells for four. Thousands of acres lie untouched; for no one attempts to work the land. They think it will yield nothing, but we know it will yield if properly cultivated. (Manuscript Release 893)

WITH GOD'S HELP 20 ACRES CAN YIELD AS MUCH AS 100: God can bless twenty acres…and make them as productive as one hundred. (Testimonies to the Church Vol. 5 p. 152)

DESERTS BLOOM AS THE GARDEN OF GOD: Even the desert places of the earth, where the outlook appears to be forbidding, may become as the garden of God. (Ministry of Healing p. 194)

MEN THOUGHT THE LAND IN COORANBONG WAS NOT GOOD, BUT GOD UNDERSTOOD THE CHEMISTRY OF THE SOIL: Before I visited Cooranbong, the Lord gave me a dream. In my dream I was taken to the land that was for sale in Cooranbong. Several of our brethren had been solicited to visit the land, and I dreamed that I was walking upon the ground. I came to a neat-cut furrow that had been plowed one quarter of a yard deep and two yards in length. Two of the brethren who had been acquainted with the rich soil of Iowa were standing before this furrow and saying, This is not good land; the soil is not favorable. But One who has often spoken in counsel was present also, and He said, False witness has been borne of this land. Then He described the properties of the different layers of earth. He explained the science of the soil, and said that this land was adapted to the growth of fruit and vegetables, and that if well worked it would produce its treasures for the benefit of man. This dream I related to Brother and Sister Starr and my family. (Manuscript Release 16 p. 153)

FAITHFUL MISSIONARIES HAVE MADE BARREN WILDERNESS PLACES BLOSSOM AS THE GARDEN OF THE LORD: From San Diego we returned to Los Angeles, and on Tuesday, December 6, we went to Redlands for a few days' visit. A little way out from Los Angeles, the scenery became very uninteresting. We passed through much barren land. Here and there, the desert, by means of irrigation, had been converted into flourishing orange groves; but for miles and miles at a stretch the land was uncultivated. As we rode along, I remembered scenes presented to me years before, of barren land, such as that through which we were passing, being cultivated and improved, and, by irrigation, made to yield rich returns. I was instructed that this was an object-lesson of the influence that the saving grace of Christ should have upon the hearts and lives of human beings. And had those to whom God has given the riches of the water of life, realized the responsibilities resting upon them as stewards of the grace of God, and gone forth as faithful missionaries into all the barren places of the earth, the wilderness would have been made to blossom as the garden of the Lord. (Review and Herald, March 30, 1905)

THINKING OUTSIDE OF THE BOX: Small fruits, such as currents, gooseberries; strawberries, raspberries, and blackberries, can be grown to advantage in many places where they are but little used and their cultivation is neglected. (Ministry of Healing p. 299)

GOD KNOWS HOW TO BRING YIELDS FROM UNPRODUCTIVE SOIL: The seed placed in thoroughly prepared soil will produce its harvest. God can spread a table for his people in the wilderness. There is much mourning over unproductive soil, when, if men would read the Old Testament Scriptures, they would see that the Lord knew much better than they in regard to the proper treatment of the land. After being worked for several years, and giving her treasures to the possession of men, portions of the land should be allowed to rest, and then the crops should be changed. We might learn much, also, from the Old Testament, in regard to the labor problem. (Advocate March 1, 1901)

THE FARM IN AUSTRALIA AN OBJECT LESSON IN WHAT COULD BE DONE WHEN OTHERS SAID NOTHING WOULD GROW: We have a large strawberry bed which will yield fruit next season. We have a few cherry trees, but the testimony is that the land is not good for cherries. But so many false, discouraging testimonies have been borne in regard to the land that we pay no attention to what they say. We shall try every kind of a tree. We have a large number of mulberry trees and fig trees of different kinds. This is not only good fruit land, but it is excellent in producing root crops and tomatoes, beans, peas, potatoes—two crops a season. All these good treasures that the land will yield have been brought in from Sydney and Newcastle and thousands of acres of land have been untouched because the owners say they will not raise anything. We have our farm as an object lesson. (Manuscript Release 8 p. 252)

— Financial Blessing #3: Self-sufficiency during the Sunday Law Crisis

AGRICULTURE TO BE A MEANS OF SUPPORT DURING THE SUNDAY LAW CRISIS: There should be land for cultivation. The time is not far distant when the laws against Sunday labor will be more stringent, and an effort should be made to secure grounds away from the cities, where fruits and vegetables can be raised. Agriculture will open resources for self-support, and various other trades also could be learned. (Fundamentals of Christian Education p. 322)

FARMING WILL BE A VIABLE MEANS OF SUPPORT WHEN THE SUNDAY LAW IS ENACTED: There should be land for cultivation. The time is not far distant when the laws against Sunday labor will be more stringent, and an effort should be made to secure grounds away from the cities, where fruits and vegetables can be raised. Agriculture will open resources for self-support, and various other trades also could be learned. This real, earnest work calls for strength of intellect as well as of muscle. Method and tact are required even to raise fruits and vegetables successfully. And habits of industry will be found an important aid to the youth in resisting temptation. (Special Testimonies on Education p. 99)

SERIOUS TIMES ARE BEFORE US: Serious times are before us, and there is great need for families to get out of the cities into the country, that the truth may be carried into the byways as well as the highways of the earth.Much depends upon laying our plans according to the word of the Lord and with persevering energy carrying them out. (Testimonies to the Church Vol. 6 p. 178)

Chapter #5: Characteristics of Model Schools that Follow God's Plan

Characteristic #1: Nature and the Bible as Primary Textbooks

GOD'S FARM AND GARDEN ARE TO BE USED AS A LESSON BOOK: The school (Avondale) is the Lord's property, and the grounds around it are His farm, where the Great Sower can make His garden a lesson book. (Testimonies to the Church Vol. 6 p. 187)

ORCHARDS AND GARDENS AS A PART OF EDUCATION: 'This land about the school is to be reserved as the school land. It is to become a living parable to the students. The students are not to regard the school land as a common thing, but are to look upon it as a lesson book open before them which the Lord would have them study. Its lessons will impart knowledge in the culture of the soul. All the land near the building is to be regarded as the school farm, where the youth can be educated under well-qualified superintendents. The youth who shall attend our schools need all the land nearby. They are to plant it with ornamental and fruit trees, and cultivate garden produce…The school farm is to be regarded as a lesson book in nature from which the teachers may draw their object lessons. Our students are to be taught that Christ, who created the world and all things that are therein, is the life and light of every living thing. The life of every child and youth who is willing to grasp the opportunities of receiving a proper education will be made thankful and happy while at school by the things upon which his eyes shall rest. (Testimonies to the Church Vol. 6 p. 181-182)

SPIRITUAL LESSONS MADE PLAIN: The land will yield its treasures, bringing the joyousness of an abundant harvest; and the produce gathered through the blessing of God is to be used as nature's lesson book, from which spiritual lessons can be made plain and applied to the necessities of the soul. (Testimonies to the Church Vol. 6 p. 187)

THE BOOK OF NATURE ONE OF THREE BOOKS FROM WHICH GOD'S WORKERS DRAW TRUTH: The great storehouse of truth is the Word of God—the written Word, the book of nature, and the book of experience in God's dealing with human life. Here are the treasures from which Christ's workers are to draw. In the search after truth they are to depend upon God, not upon human intelligences, the great men whose wisdom is foolishness with God. Through His own appointed channels the Lord will impart a knowledge of Himself to every seeker. But it is in the written Word that a knowledge of God is most clearly revealed to fallen man. This is the treasure house of the unsearchable riches of Christ.... The truths of redemption are capable of constant development and expansion. Though old, they are ever new, constantly revealing to the seeker for truth a greater glory and a mightier power (Christ's Object Lessons, pp. 125-127).

THE BIBLE AND THE BOOK OF NATURE SHED LIGHT ON EACH OTHER: The book of nature and the written Word shed light upon each other. Both make us better acquainted with God by teaching us of His character and of the laws through which He works. (God's Amazing Grace p. 286)

THE BIBLE TOUCHES ON NEARLY EVERY OPERATION OF NATURE: There is scarcely an operation of nature to which we may not find reference in the Word of God. The Word declares that he maketh his sun to rise, and the rain to descend. He maketh grass to grow upon the mountains.... He giveth snow like wool: he scattereth the hoarfrost like ashes. He casteth forth his ice like morsels.... He sendeth out his word, and melteth them: he causeth his wind to blow, and the waters flow. He maketh lightnings for the rain; and bringeth the wind out of his treasuries. (To Be Like Jesus p. 237)

LESSONS DRAWN OUT WHILE WORKING: Let the children themselves prepare the soil and sow the seed. As they work, the parent or teacher can explain the garden of the heart with the good or bad seed sown there, and that as the garden must be prepared for the natural seed, so the heart must be prepared for the seed of truth. As the seed is cast into the ground, they can teach the lesson of Christ's death; and as the blade springs up, they can teach the lesson of the truth of the resurrection. As the plants grow, the correspondence between the natural and the spiritual sowing may be continued. The youth should be instructed in a similar way. They should be taught to till the soil. It would be well if there were connected with every school, lands for cultivation. Such lands should be regarded as God's own schoolroom. The things of nature should be looked upon as a lesson book which His children are to study, and from which they may obtain knowledge as to the culture of the soul. (Christ's Object Lessons p. 87-88)

QUESTIONS TO SET MINDS THINKING: While the children and youth gain a knowledge of facts from teachers and textbooks, let them learn to draw lessons and discern truth for themselves. In their gardening, question them as to what they learn from the care of their plants. When they gather the flowers, lead them to think why He spared us the beauty of these wanderers from Eden. Teach them to notice the evidences everywhere manifested in nature of God's thought for us, the wonderful adaptation of all things to our need and happiness. (Education p. 119)

OPPORTUNITIES FOR STUDENTS TO SHARE: Students in the industrial departments, whether they are employed in domestic work, in cultivating the ground, or in other ways, should have time and opportunity given them to tell the practical, spiritual lessons they have learned in connection with the work. (Testimonies to the Church Vol. 6 p. 177)

NATURE THE TEXTBOOK IN EDEN'S SCHOOL: The whole natural world is designed to be an interpreter of the things of God. To Adam and Eve in their Eden home, nature was full of the knowledge of God, teeming with divine instruction. To their attentive ears it was vocal with the voice of wisdom. Wisdom spoke to the eye and was received into the heart, for they communed with God in His created works. The book of nature, which spread its living lessons before them, afforded an exhaustless source of instruction and delight. On every leaf of the forest and stone of the mountains, in every shining star, in earth and sea and sky, God's name was written. With both the animate and the inanimate creation—with leaf and flower and tree, and with every living creature, from the leviathan of the waters to the mote in the sunbeam—the dwellers in Eden held converse, gathering from each the secrets of its life. God's glory in the heavens, the innumerable worlds in their orderly revolutions, the balancings of the clouds (Job 37:16), the mysteries of light and sound, of day and night—all were objects of study by the pupils of earth's first school. (Child Guidance p. 45)

NATURE WAS ADAM AND EVE'S LESSON BOOK, GOD THEIR TEACHER: Before the fall of Adam, not a cloud rested on the mind of our first parents to obscure their clear perception of the divine character of God. They were perfectly conformed to the will of God. A beautiful light, the light of God, surrounded them. Nature was their lesson book. The Lord instructed them in regard to the natural world and then left with them this open book that they might behold beauty in every object upon which their eyes should rest. The Lord visited the holy pair, and instructed them through the works of His hands. (The Upward Look p. 198)

AFTER THE FALL NATURE COULD NO LONGER BE THE ONLY TEACHER OF MAN: The beauties of nature are an expression of the love of God for human intelligences, and in the Garden of Eden the existence of God was demonstrated in the objects of nature that surrounded our first parents. Every tree planted in the Garden spoke to them, saying that the invisible things of God were clearly seen, being understood by the things which were made, even His eternal power and Godhead. But while thus God could be discerned in nature, this affords no solid argument in favor of a perfect knowledge of God being revealed in nature to Adam and his posterity after the Fall. Nature could convey her lessons to man in his innocence, but sin and transgression brought a blight upon nature, and intervened between nature and nature's God. Had man never disobeyed his Creator, had he remained in his state of perfect rectitude, he could have understood and known God. But when man disobeyed God, he gave evidence that he believed the words of an apostate rather than the words of God.... Adam and Eve listened to the voice of the tempter, and sinned against God. The light, the garments of heavenly innocence, departed from these tried, deceived souls, and in parting with the garments of innocence, they drew about them the dark robes of ignorance of God. The clear and perfect light of innocence, which had hitherto surrounded them, had lightened everything which they approached; but deprived of that heavenly light, the posterity of Adam could no longer trace the character of God in His created works. Therefore, after the Fall, nature was not the only teacher of man. In order that the world might not remain in darkness, in eternal, spiritual night, the God of nature must meet man in Jesus Christ. The Son of God came to the world as a revelation of the Father. He was the true light, which lighteth every man that cometh into the world. (Manuscript 86, July 3, 1898, Notes of the Week of Prayer)

THROUGH CREATION STUDENTS BECOME ACQUAINTED WITH GOD: So through the creation we are to become acquainted with the Creator. The book of nature is a great lesson book, which in connection with the Scriptures we are to use in teaching others of His character, and guiding lost sheep back to the fold of God. As the works of God are studied, the Holy Spirit flashes conviction into the mind. It is not the conviction that logical reasoning produces; but unless the mind has become too dark to know God, the eye too dim to see Him, the ear too dull to hear His voice, a deeper meaning is grasped, and the sublime, spiritual truths of the written word are impressed on the heart. In these lessons direct from nature, there is a simplicity and purity that makes them of the highest value. All need the teaching to be derived from this source. (Christ's Object Lessons p. 24)

JESUS ILLUSTRATED IMMORTAL TRUTHS WITH LESSONS FROM NATURE: The birds of the air, the lilies of the field, the sower and the seed, the shepherd and the sheep—with these Christ illustrated immortal truth. He drew illustrations also from the events of life, facts of experience familiar to the hearers—the leaven, the hid treasure, the pearl, the fishing net, the lost coin, the prodigal son, the houses on the rock and the sand. In His lessons there was something to interest every mind, to appeal to every heart. Thus the daily task, instead of being a mere round of toil, bereft of higher thoughts, was brightened and uplifted by constant reminders of the spiritual and the unseen. So we should teach. Let the children learn to see in nature an expression of the love and the wisdom of God; let the thought of Him be linked with bird and flower and tree; let all things seen become to them the interpreters of the unseen, and all the events of life be a means of divine teaching. As they learn thus to study the lessons in all created things, and in all life's experiences, show that the same laws which govern the things of nature and the events of life are to control us; that they are given for our good; and that only in obedience to them can we find true happiness and success. (Education pp. 102-103)

NATURE ILLUSTRATES BIBLE LESSONS: Many illustrations from nature are used by the Bible writers; and as we observe the things of the natural world, we shall be enabled, under the guiding of the Holy Spirit, more fully to understand the lessons of God's Word. In the natural world God has placed in the hands of the children of men the key to unlock the treasure house of His Word. The unseen is illustrated by the seen; divine wisdom, eternal truth, infinite grace, are understood by the things that God has made. Children should be encouraged to search out in nature the objects that illustrate Bible teachings, and to trace in the Bible the similitudes drawn from nature. They should search out, both in nature and in Holy Writ, every object representing Christ, and those also that He employed in illustrating truth. Thus may they learn to see Him in tree and vine, in lily and rose, in sun and star. They may learn to hear His voice in the song of birds, in the sighing of the trees, in the rolling thunder, and in the music of the sea. And every object in nature will repeat to them His precious lessons. To those who thus acquaint themselves with Christ, the earth will nevermore be a lonely and desolate place. It will be their Father's house, filled with the presence of Him who once dwelt among men. (Child Guidance pp. 46-47)

NATURE ILLUMINATES SPIRITUAL TRUTH: If the follower of Christ will believe His word and practice it, there is no science in the natural world that he will not be able to grasp and appreciate. There is nothing but that will furnish him means for imparting the truth to others. Natural science is a treasure house of knowledge from which every student in the school of Christ may draw. As we contemplate the beauty of nature, as we study its lessons in the cultivation of the soil, in the growth of the trees, in all the wonders of earth and sea and sky, there will come to us a new perception of truth. And the mysteries connected with God's dealings with men, the depths of His wisdom and judgment as seen in human life—these are found to be a storehouse rich in treasure. (Christ's Object Lessons p. 125)

LIGHT FROM CAVALRY INTERPRETS NATURE ARIGHT: Only in the light that shines from Calvary can nature's teaching be read aright. Through the story of Bethlehem and the cross let it be shown how good is to conquer evil, and how every blessing that comes to us is a gift of redemption. In brier and thorn, in thistle and tare, is represented the evil that blights and mars. In singing bird and opening blossom, in rain and sunshine, in summer breeze and gentle dew, in ten thousand objects in nature, from the oak of the forest to the violet that blossoms at its root, is seen the love that restores. And nature still speaks to us of God's goodness. As the dwellers in Eden learned from nature's pages, as Moses discerned God's handwriting on the Arabian plains and mountains, and the Child Jesus on the hillsides of Nazareth, so the children of today may learn of Him. The unseen is illustrated by the seen. On everything upon the earth ... from the boundless ocean to the tiniest shell on the shore, they may behold the image and superscription of God (Child Guidance p. 47)

FOLLOWING JESUS' EXAMPLE IN STUDYING FROM THE SCRIPTURES AND NATURE: [Jesus'] education was gained from Heaven-appointed sources, from useful work, from the study of the Scriptures, from nature, and from the experiences of life—God's lesson books, full of instruction to all who bring to them the willing hand, the seeing eye, and the understanding heart. His intimate acquaintance with the Scriptures shows how diligently His early years were given to the study of God's Word. And spread out before Him was the great library of God's created works. He who had made all things studied the lessons which His own hand had written in earth and sea and sky. Apart from the unholy ways of the world, He gathered stores of scientific knowledge from nature. He studied the life of plants and animals, and the life of man. From His earliest years He was possessed of one purpose; He lived to bless others. For this He found resources in nature; new ideas of ways and means flashed into His mind as He studied plant life and animal life....Thus to Jesus the significance of the Word and the works of God was unfolded, as He was trying to understand the reason of things. Heavenly beings were His attendants, and the culture of holy thoughts and communings was His. From the first dawning of intelligence He was constantly growing in spiritual grace and knowledge of truth. Every child may gain knowledge as Jesus did. As we try to become acquainted with our heavenly Father through His Word, angels will draw near, our minds will be strengthened, our characters will be elevated and refined. (Child Guidance p. 50)

Characteristic #2: Hand-in-Hand Integration with an Agricultural Program

AGRICULTURAL WORK THE GRANDEST, MOST HELPFUL ALL-AROUND EDUCATION THE YOUTH CAN HAVE: This is the grandest, the most helpful, all-around education that the youth can have. Cultivating the soil, planting and caring for trees, sowing seed and watching its growth,—this work teaches precious lessons. Nature is an expositor of the word of the living God. But only through Christ does creation answer the highest purpose of the Creator. The Savior has wonderful revelations for all who will walk humbly with God. Under the discipline and training of the higher teaching, they will behold wondrous things out of His law. (Manuscript Release 6 p. 410)

AGRICULTURE THE ABC OF INDUSTRIAL EDUCATION: At the Huntsville School [Oakwood] a thorough work is to be done in training men to cultivate the soil and to grow fruits and vegetables. Let no one despise this work. Agriculture is the ABC of industrial education. Let the erection of the buildings for the school and the sanitarium be an education to the students. Help the teachers to understand that their perceptions must be clear, their actions in harmony with the truth, for it is only when they stand in right relation to God that they will be able to work out His plan for themselves and for the souls with whom, as instructors, they are brought in contact. (A Place Called Oakwood p. 45)

STUDY OF TEXTBOOKS ALONE CANNOT ACCOMPLISH THE WORK: In our schools the standard of education must not be lowered. It must be lifted higher and still higher, far above where it now stands; but the education given must not be confined to a knowledge of textbooks merely. The study of textbooks alone cannot afford students the discipline they need, nor can it impart true wisdom. The object of our schools is to provide places where the younger members of the Lord's family may be trained according to His plan of growth and development. (Testimonies to the Church Vol. 6 pp. 126-127)

MORE EFFORT SHOULD BE PUT INTO TEACHING STUDENTS ABOUT AGRICULTURE: A greater effort should be made to create and to encourage an interest in agricultural pursuits. Let the teacher call attention to what the Bible says about agriculture: that it was God's plan for man to till the earth; that the first man, the ruler of the whole world, was given a garden to cultivate; and that many of the world's greatest men, its real nobility, have been tillers of the soil. Show the opportunities in such a life. (Education p. 218)

STUDY ALONG AGRICULTURAL LINES SHOULD BE THE A, B, AND C OF THE EDUCATIONAL WORK: From the light given me in regard to the location and building up of our school interests, I know that it is the purpose of God that this institution be established at a distance from the city that is so full of temptations and snares, of amusements and holidays, which are not conducive to purity and piety and religious devotion. He designs that we shall connect manual labor with the improvement of the mental powers. I have been shown that study in agricultural lines should be the A B and C of the educational work of our schools. (Spalding and Magan p. 134)

STUDY IN AGRICULTURAL LINES TO BE THE A, B, AND C OF CHRISTIAN EDUCATION—THE VERY FIRST WORK THAT SHOULD BE ENTERED UPON: Working the soil is one of the best kinds of employment, calling the muscles into action and resting the mind. Study in agricultural lines should be the A, B, and C of the education given in our schools. This is the very first work that should be entered upon. Our schools should not depend upon imported produce, for grain and vegetables, and the fruits so essential to health. Our youth need an education in felling trees and tilling the soil as well as in literary lines. Different teachers should be appointed to oversee a number of students in their work and should work with them. Thus the teachers themselves will learn to carry responsibilities as burden bearers. Proper students also should in this way be educated to bear responsibilities and to be laborers together with the teachers. All should counsel together as to the very best methods of carrying on the work. (Testimonies to the Church Vol. 6 p. 179)

THE AGRICULTURAL COMPONENT SHOULD BE THE FIRST WORK ENTERED UPON IN STARTING A SCHOOL: This institution must not depend upon imported produce,—for the fruits so essential to healthfulness, and for their grains and vegetables. This is the very first work that must be entered upon. Then as we shall advance and add to our facilities, advance studies and object lessons should come in. We are not to subtract from that which has already been taken hold of as a branch of education. (Spalding and Magan p. 134)

SCHOOLS TO BE SMALL AND CONNECTED TO FARMS: The Lord permitted fire to consume the principal buildings of the Review and Herald and the sanitarium, and thus removed the greatest objection urged against moving out of Battle Creek. It was His design that instead of rebuilding the one large sanitarium, our people should make plants in several places. These smaller sanitariums should have been established where land could be secured for agricultural purposes. It is God's plan that agriculture shall be connected with the work of our sanitariums and schools. Our youth need the education to be gained from this line of work. It is well, and more than well—it is essential—that efforts be made to carry out the Lord's plan in this respect. (Counsels on Health p. 223)

FARMING AND INDUSTRY SHOULD BE CONNECTED WITH SDA COLLEGES: It would be well could there be connected with our college, land for cultivation and also workshops under the charge of men competent to instruct the students in the various departments of physical labor. Much is lost by a neglect to unite physical with mental taxation. The leisure hours of the students are often occupied with frivolous pleasures, which weaken physical, mental, and moral powers. Under the debasing power of sensual indulgence, or the untimely excitement of courtship and marriage, many students fail to reach that height of mental development which they might otherwise have attained. (Testimonies to the Church Vol. 5 p. 23)

THE FARM IN COORANBONG WAS AN OBJECT LESSON, TOTALLY WRAPPED UP IN THE EDUCATION OF THE STUDENTS: We must open to our students the Book of all books, the living oracles of God. Here true wisdom is to be found. In all matters that pertain to our present duty to God, and to our future, eternal interests, we may here receive divine instruction. And we are to learn from nature. We thank the Lord that we are located just where we are. The land we are cultivating as the school farm is testifying to all that false witness has been borne against it. We are making this land an educating book for the students. From it they are to learn the meaning of the words, We are laborers together with God; ye are God's husbandry; ye are God's building. (Australasian Union Conference Record, January 1, 1900)

FAMILIES AND INSTITUTIONS NEED TO DO MORE IN CULTIVATING THE LAND: Families and institutions should learn to do more in the cultivation and improvement of land. If people only knew the value of the products of the ground, which the earth brings forth in their season, more diligent efforts would be made to cultivate the soil. All should be acquainted with the special value of fruit and vegetables fresh from the orchard and garden. (Counsels on Diet and Foods p. 312)

AN EMPHASIS ON AGRICULTURAL LINES WOULD HAVE PREPARED STUDENTS FOR THE LORD'S FARM IN THE EARTH MADE NEW. DIFFICULTIES WOULD HAVE BEEN OVERCOME, AND THE SCHOOLS WOULD BE MUCH MORE SUCCESSFUL: Let the students be out in the most healthful location that can be secured, to do the very work that should have been done years ago. Then there would not be so great discouragements. Had this been done, you would have had some grumbling from students, and many objections would have been raised by parents, but this all-round education would educate the children and youth, not only for practical work in various trades, but would prepare them for the Lord's farm in the earth made new. If all in America had encouraged the work in agricultural lines that principals and teachers have discouraged, the schools would have had altogether a different showing. Opposing influences would have been overcome; circumstances would have changed; there would have been greater physical and mental strength; labor would have been equalized; and the taxing of all the human machinery would have proved the sum. (Manuscript Release 8 p. 199)

TRUE EDUCATION INVOLVES MORE THAN BOOK KNOWLEDGE: To give students a knowledge of books merely is not the purpose of the institution. Such education can be obtained at any college in the land. I was shown that it is Satan's purpose to prevent the attainment of the very object for which the college was established. Hindered by his devices, its managers reason after the manner of the world and copy its plans and imitate its customs. But in thus doing, they will not meet the mind of the Spirit of God. (Testimonies to the Church Vol. 5 p. 22)

GOD'S PLAN CALLS FOR FARMING TO BE CONNECTED WITH OUR SCHOOLS: It is God's plan that agriculture shall be carried on in connection with our sanitariums and schools. Our youth need the education to be gained from this line of work. It is well, and more than well—it is essential—that efforts be made to carry out the Lord's plan in this respect. (Manuscript Release 11 p. 42)

SCHOOLS FOR THE LAST DAYS TO BE OF AN ENTIRELY DIFFERENT ORDER THAN THOSE OF THE WORLD: God has revealed to me that we are in positive danger of bringing into our educational work the customs and fashions that prevail in the schools of the world. If teachers are not guarded, they will place on the necks of their students worldly yokes instead of the yoke of Christ. The plan of the schools we shall establish in these closing years of the message is to be of an entirely different order from those we have instituted. (Counsels to Parents, Teachers and Students p. 532)

SCHOOLS IN THE LAST DAYS TO BE POSITIVELY DIFFERENT FROM THOSE ESTABLISHED IN THE PAST: God has revealed to me that we are in positive danger of bringing into our educational work the customs and fashions that prevail in the schools of the world. If teachers are not guarded in their work, they will place on the necks of their students worldly yokes instead of the yoke of Christ. The plan of the schools we shall establish in these closing years of the work is to be of an entirely different order from those we have instituted in the past. (Counsels to Parents, Teachers, and Students p. 532)

SCHOOLS SHOULD BE ESTABLISHED WHERE A KNOWLEDGE OF PRACTICAL INDUSTRY—AS WELL AS BOOKS—IS TAUGHT. RICHES CAN BE OBTAINED FROM THE SOIL, BUT INSTRUCTION IS NEEDED. Australia needs the leaven of sound, solid, common sense to be freely introduced into all her cities and towns. There is need of proper education. Schools should be established for the purpose of obtaining not only knowledge from books, but knowledge of practical industry. Men are needed in different communities to show the people how riches are to be obtained from the soil. The cultivation of land will bring its return. (Special Testimonies on Education p. 92)

THROUGH AGRICULTURAL TRAINING, MANY STUDENTS WILL LEARN TO LOVE FARMING, BE INFLUENCED IN THEIR CHOICE OF OCCUPATION, AND GRAVITATE TOWARDS COUNTRY LIVING: In the study of agriculture, let pupils be given not only theory, but practice. While they learn what science can teach in regard to the nature and preparation of the soil, the value of different crops, and the best methods of production, let them put their knowledge to use. Let teachers share the work with the students, and show what results can be achieved through skillful, intelligent effort. Thus may be awakened a genuine interest, an ambition to do the work in the best possible manner. Such an ambition, together with the invigorating effect of exercise, sunshine, and pure air, will create a love for agricultural labor that with many youth will determine their choice of an occupation. Thus might be set on foot influences that would go far in turning the tide of migration which now sets so strongly toward the great cities. (Education p. 219)

YOUTH SHOULD BE TRAINED TO LOVE WORKING ON THE LAND: We should so train the youth that they will love to work upon the land, and delight in improving it. (Special Testimonies on Education p. 99)

EDUCATION REGARDING PREPARATION OF THE SOIL IS VERY MUCH NEEDED. Poverty and distress are on every hand. There are men who work hard, and obtain very little for their labor. There is need of much more extensive knowledge in regard to the preparation of the soil. There is not sufficient breadth of view as to what can be realized from the earth. A narrow and unvarying routine is followed with discouraging results. (Special Testimonies on Education p. 93)

Chapter #5: Characteristics of Model Schools that Follow God's Plan

EDUCATION SHOULD BE MORE COMPREHENSIVE: A more comprehensive education is needed, an education which will demand from teachers and principal such thought and effort as mere instruction in the sciences does not require. The character must receive proper discipline for its fullest and noblest development. The students should receive at college such training as will enable them to maintain a respectable, honest, virtuous standing in society, against the demoralizing influences which are corrupting the youth. (Testimonies to the Church Vol. 5 p. 23)

EVERY PURSUIT OF THE CHRISTIAN—INCLUDING AGRICULTURAL INTERESTS—TO BE BOUND UP WITH HEAVENLY THINGS: The thought of the eternal life should be woven into all to which the Christian sets his hand. If the work performed is agricultural or mechanical in its nature, it may still be after the pattern of the heavenly. It is the privilege of the preceptors and teachers of our schools to reveal in all their work the leading of the Spirit of God. Through the grace of Christ every provision has been made for the perfecting of Christlike characters; and God is honored when His people, in all their social and business dealings, reveal the principles of heaven. (Counsels to Parents, Teachers and Students p. 58)

Characteristic #3: Balance between Mental and Physical Labor

BALANCE BETWEEN PHYSICAL AND MENTAL TAXATION BEST FOR BOTH CHILDREN AND MEN: For both children and men, labor combined with mental taxation will give the right kind of all-round education. The cultivation of the mind will bring tact and fresh incentive to the cultivation of the soil. (Manuscript Release 11 p. 184)

TRUE EDUCATION INVOLVES MORE THAN A KNOWLEDGE OF BOOKS: Education comprises more than a knowledge of books. Proper education includes not only mental discipline, but that training which will secure sound morals and correct deportment. (Testimonies to the Church Vol. 4 p. 648)

ACADEMIC STUDY ALONE CANNOT GIVE STUDENTS WHAT THEY NEED: Our college is designed of God to meet the advancing wants for this time of peril and demoralization. The study of books only cannot give students the discipline they need. A broader foundation must be laid. The college was not brought into existence to bear the stamp of any one man's mind. Teachers and principal should work together as brethren. They should consult together, and also counsel with ministers and responsible men, and, above all else, seek wisdom from above, that all their decisions in reference to the school may be such as will be approved of God. (Testimonies to the Church Vol. 5 p. 22)

EQUAL TAXATION OF BRAIN AND MUSCLE NEEDED FOR GOOD EDUCATION: The close confinement of students to mental work has cost the life of many precious youth. The Madison school, in its system of education, is showing that mental and physical powers, brain and muscle, must be equally taxed. The example that it has given in this respect is one that it would be well for all who engage in school work to emulate. If the physical and mental powers were equally taxed, there would be in our world far less of corruption of mind and far less feebleness of health. (Letter 168, 1908 to J. E. White and wife)

A PORTION OF EACH DAY TO BE DEVOTED TO LABOR: In connection with the schools should have been agricultural and manufacturing establishments…A portion of each day should have been devoted to labor, that the physical and mental powers might be equally exercised. (Testimonies to the Church Vol. 3 p. 153)

OVEREMPHASIS ON STUDY AND THE ACADEMIC LEADS TO UNBALANCED MINDS AND UNFITS YOUTH FOR PRACTICAL LIFE. IN ORDER TO PRESERVE BALANCED MINDS, LABOR AND STUDY SHOULD BE UNITED IN SCHOOLS: All the powers of the mind should be called into use, and developed, in order for men and women to have well-balanced minds. The world is full of one-sided men and women, because one set of the faculties is cultivated, while others are dwarfed from inaction. The education of most youth is a failure. They overstudy, while they neglect that which pertains to practical business life. Men and women become parents without considering their responsibilities, and their offspring sink lower in the scale of human deficiency than they themselves. Thus we are fast degenerating. The constant application to study, as the schools are now conducted, is unfitting youth for practical life. The human mind will have action. If it is not active in the right direction, it will be active in the wrong. And in order to preserve the balance of the mind, labor and study should be united in schools. (Christian Education p. 17)

ACADEMIC LOADS SHOULD NOT BE SET BY THE STUDENTS. THERE SHOULD BE LESS STUDY OF BOOKS AND MORE PRACTICAL LESSONS IN THE TRADES: All the arts are to come into the education of the students. Even in the school at Avondale there are too many studies taken by the students. The youth should not be left to take all the studies they shall choose, for they will be inclined to take more than they can carry; and if they do this, they cannot possibly come from the school with a thorough knowledge of each study. There should be less study of books, and greater painstaking effort made to obtain that knowledge which is essential for practical life. The youth are to learn how to work interestedly and intelligently, that, wherever they are, they may be respected because they have a knowledge of those arts which are so essential for practical life. In the place of being day laborers under an overseer, they are to strive to be masters of their trades to place themselves where they can command wages as good carpenters, printers, or as educators in agricultural work. (Manuscript 105, 1898, pp. 2-3)

THE BODY AND BRAIN SHOULD BE EQUALLY TAXED: The most astonishing ignorance prevails in regard to putting brain, bone, and muscle into active service. Every part of the human organism should be equally taxed. This is necessary for the harmonious development and action of every part. (Manuscript Release 19 p. 230)

AN OVEREMPHASIS ON WORK IS ALSO UNHEALTHY—THERE MUST BE A BALANCE: Dear Brother E, you have made a great mistake in giving this world your ambition. You are exacting and sometimes impatient, and at times require too much of your son. He has become discouraged. At your house it has been work, work, work, from early morning until night. Your large farm has brought extra cares and burdens into your house. You have talked upon business; for business was primary in your mind, and out of the abundance of the heart the mouth speaketh. Has your example in your family exalted Christ and His salvation above your farming interest and your desire for gain? If your children fail of everlasting life, the blood of their souls will surely be found on the garments of their father. (Testimonies to the Church Vol. 4 p. 48)

THE IMPORTANCE OF BALANCE IN PHYSICAL AND MENTAL LABOR: We are to educate the youth to exercise equally the mental and the physical powers. The healthful exercise of the whole being will give an education that is broad and comprehensive. We had stern work to do in Australia in educating parents and youth along these lines; but we persevered in our efforts until the lesson was learned that in order to have an education that was complete, the time of study must be divided between the gaining of book knowledge and the securing of a knowledge of practical work. Part of each day was spent in useful work, the students learning how to clear the land, how to cultivate the soil, and how to build houses, using time that would otherwise have been spent in playing games and seeking amusement. And the Lord blessed the students who thus devoted their time to acquiring habits of usefulness. (Review and Herald, November 11, 1909)

WARNING AGAINST OVERWORK: May the peace of God abide in your home. May His blessing rest upon your little flock. They are lambs of His fold, and must be nurtured and cherished. Do not overwork. Do not strain every nerve and muscle toto try to do everything that there is to do on the farm, but get help. (Manuscript Release 16 p. 186)

— (1): Manual Labor an Integral Part of Curriculum

And whatsoever ye do, do it heartily, as to the Lord, and not unto men; knowing that of the Lord ye shall receive the reward of the inheritance: for ye serve the Lord Christ. COLOSSIANS 3:23-24

A MUCH BIGGER PLAN FOR OUR SCHOOLS WHICH INVOLVES TRADES AND MANUAL LABOR: Our schools must be conducted under the supervision of God. There is a work to be done for young men and women that is not yet accomplished. There are much larger numbers of young people who need to have the advantages of our training schools. They need the manual training course, that will teach them how to lead an active, energetic life. All kinds of labor must be connected with our schools. Under wise, judicious, God-fearing directors the students are to be taught. Every branch of the work is to be conducted in the most thorough and systematic ways that long experience and wisdom can enable us to plan and execute. (Testimonies to the Church Vol. 6 p. 191)

MANUAL TRAINING TO BE CONNECTED WITH EVERY SCHOOL: So far as possible, facilities for manual training should be connected with every school. To a great degree such, training would supply the place of the gymnasium, with the additional benefit of affording valuable discipline. (Education p. 217)

LEARNING THE PRACTICAL DUTIES OF LIFE: Some hours each day should be devoted to useful education of work that will help the students in learning the duties of practical life, which are essential for all our youth (Counsels to Parents, Teachers and Students p. 283)

AVOIDING A DEADLY IMPRESSION: The impression that work is degrading has laid thousands in the grave. Those of sedentary and literary habits should take physical exercise. Health should be a sufficient inducement to lead them to unite physical with their mental labor. (Testimonies to the Church Vol. 6 p. 192)

DANGERS OF SEDENTARY OCCUPATIONS: …the occupations requiring sedentary habits are the most dangerous, for they take men away from the open air and sunshine. Men carry on their work, perfect their business, and soon lie down in the grave. Much more favorable is the condition of one whose occupation keeps him in the open air, exercising his muscles, while the brain is equally taxed, and all the organs have the privilege of doing their work. (Fundamentals of Christian Education p. 319)

SYSTEMATIC, WELL-REGULATED LABOR NEEDED FOR ALL-ROUND DEVELOPMENT: The youth need a training that will make them practical—men and women who can cope with emergencies. They should be taught that the discipline of systematic, well-regulated labor is essential, not only as a safeguard against the vicissitudes of life, but as an aid to all-round development. (Child Guidance p. 347)

DISCIPLINE OF SYSTEMATIC, WELL-REGULATED LABOR ESSENTIAL TO ALL-ROUND DEVELOPMENT: The youth need to be taught that life means earnest work, responsibility, caretaking. They need a training that will make them practical—men and women who can cope with emergencies. They should be taught that the discipline of systematic, well-regulated labor is essential, not only as a safeguard against the vicissitudes of life, but as an aid to all-around development. (To Be Like Jesus p. 234)

ALL STUDENTS SHOULD SPEND A PORTION OF EACH DAY IN PHYSICAL LABOR: There should have been in past generations provisions made for education upon a larger scale. In connection with the schools should have been agricultural and manufacturing establishments. There should have been teachers also of household labor. There should have been a portion of the time each day devoted to labor, that the physical and mental might be equally exercised. If schools had been established upon the plan we have mentioned, there would not now be so many unbalanced minds. (Christian Education p. 18)

EVERY INSTITUTION OF LEARNING TO PROVIDE FOR PRACTICAL TRAINING: Every institution of learning should make provision for the study and practice of agriculture and the mechanic arts. Competent teachers should be employed to instruct the youth in the various industrial pursuits, as well as in the several branches of study. While a part of each day is devoted to mental improvement, let a stated portion be given to physical labor, and a suitable time to devotional exercises and the study of the Scriptures. (Fundamentals of Christian Education p. 72-73)

PHYSICAL LABOR IS PART OF GOD'S PLAN FOR EVERY MAN: The school to be established in Australia should bring the question of industry to the front, and reveal the fact that physical labor has its place in God's plan for every man, and that his blessing will attend it. The schools established by those who teach and practise the truth for this time, should be so conducted as to bring fresh and new incentives into all kinds of practical labor. There will be much to try the educators, but a great and noble object has been gained when students shall feel that love for God is to be revealed, not only in the devotion of heart and mind and soul, but in the apt, wise appropriation of their strength. Their temptations will be far less; from them by precept and example a light will radiate amid the erroneous theories and fashionable customs of the world. Their influence will tend to correct the false idea that ignorance is the mark of a gentleman. (Special Testimonies on Education p. 101)

LEARN TO LOVE USEFUL ACTIVITY: The habit of enjoying useful labor, once formed, will never be lost. You are then prepared to be placed in any circumstance in life, and you will be fitted for the position. You will learn to love activity. (Testimonies to the Church Vol. 3 p. 336)

DAILY SYSTEMATIC LABOR TO BE PART OF EDUCATION: Daily systematic labor should constitute a part of the education of the youth. In following this plan the students will realize elasticity of spirit and vigor of thought, and will be able to accomplish more mental labor

Chapter #5: Characteristics of Model Schools that Follow God's Plan

in a given time than they could by study alone. (Counsels to Parents, Teachers and Students p. 292)

THE IMPORTANCE OF REGULAR MANUAL LABOR—INCLUDING FARMING—IN THE WORK OF OUR SCHOOLS: Provision should have been made in past generations for education upon a larger scale. In connection with the schools should have been agricultural and manufacturing establishments. There should also have been teachers of household labor. And a portion of the time each day should have been devoted to labor that the physical and mental powers might be equally exercised. If schools had been established on the plan we have mentioned, there would not now be so many unbalanced minds… (Counsels to Parents, Teachers and Students p. 288)

EMPHASIS NEEDED IN AGRICULTURAL AND PRACTIAL LINES: The Lord calls for steps in advance. Because the teachers may never have been trained to physical, manual labor, they are not easily persuaded in regard to the very best methods to secure for the youth an all-round education, and even the very ones who have been most reluctant to come into line in this matter, had they been given in their youth the physical, mental, and moral education combined, might have saved themselves several attacks of illness, and their brain, bone, and muscle would at this time be in a more healthy condition because all of the Lord's machine would be proportionately taxed. Precious lessons from the best instructors should be secured in spiritual lines, in agricultural employments, and also in carpenter's trade and in the printing business. The Lord would have these mechanical industries brought in and taught by competent men. (Spalding and Magan p. 204)

MULTIPLE LINES OF PRACTICAL LABOR TO BE CONNECTED WITH OUR SCHOOLS: Our schools must be conducted under the supervision of God. There is a work to be done for young men and young women that is not yet accomplished. There are much larger numbers of young people who need to have the advantages of our school. They need the manual-training course, which will teach them how to lead an active, energetic life. All kinds of labor must be connected with our school. Under wise, judicious, God-fearing directors, the students are to be taught. Every branch of the work is to be conducted on the most thorough and systematic lines that long experience and wisdom can plan and execute. (Advocate, February 1, 1899)

STUDENTS SHOULD LEARN FARMING, MASONRY, BUILDING, TENT MAKING, AND PRINTING: The education to be gained in the felling of trees, the tilling of the soil, and the erection of buildings, as well as the studies of the classroom, is what our youth should seek to obtain. Tent making also should be taught, buildings should be erected, and masonry should be learned. Further on, a printing press should be connected with the school, that an education may be given to students in this line of work. (Manuscript Release 8 p. 152)

ALL STUDENTS—WHETHER RICH OR POOR—SHOULD LEARN TO CULTIVATE THE LAND AND HAVE A BALANCE BETWEEN PHYSICAL AND MENTAL TRAINING: There should be work for all students, whether they are able to pay their way or not; the physical and mental powers should receive proportionate attention. Students should learn to cultivate the land; for this will bring them into close contact with nature. (Special Testimonies on Education p. 46)

DAILY, SYSTEMATIC LABOR NEEDFUL FOR STUDENTS TO GAIN ELASTICITY OF SPIRIT AND VIGOR OF THOUGHT: Time is too short now to accomplish that which might have been done in past generations. But even in these last days we can do much to correct the existing evils in the education of youth. And because time is short, we should be in earnest and work zealously to give the young an education consistent with our faith. We are reformers. We desire that our children should study to the best advantage. In order to do this, employment should be given them which will call into exercise the muscles. Daily, systematic labor should constitute a part of the education of youth even at this late period. Much can now be gained in this way. In following this plan the students will realize elasticity of spirit and vigor of thought, and in a given time can accomplish more mental labor than they could by study alone. And thus they can leave school with constitutions unimpaired and with strength and courage to persevere in any position where the providence of God may place them. (Testimonies to the Church Vol. 6 p. 179)

PRACTICAL TRADES—INCLUDING AGRICULTURE—PART OF GOD'S EDUCATIONAL PLAN FOR THE YOUTH: Carpentering, blacksmithing, agriculture, the best way to make the most of what the earth produces—all these things are part of the education to be given to the youth. (Letter 25, 1902)

SKILL ALONG PRACTICAL LINES A VALUABLE GIFT FROM GOD: Skill in the common arts is a gift from God. He provides both the gift and wisdom to use the gift aright. When He desired a work done on the tabernacle He said, See, I have called by name Bezaleel the son of Uri, the son of Hur, of the tribe of Judah: and I have filled him with the Spirit of God, in wisdom, and in understanding, and in knowledge, and in all manner of workmanship. Exodus 31:2, 3. Through the prophet Isaiah the Lord said, Give ye ear, and hear My voice; hearken, and hear My speech. Doth the plowman plow all day to sow? doth he open and break the clods of his ground? When he hath made plain the face thereof, doth he not cast abroad the fitches, and scatter the cumin, and cast in the principal wheat and the appointed barley and the rye in their place? For his God doth instruct him to discretion, and doth teach him. For the fitches are not threshed with a threshing instrument, neither is a cart wheel turned about upon the cumin; but the fitches are beaten out with a staff, and the cumin with a rod. Bread corn is bruised; because he will not ever be threshing it, nor break it with the wheel of his cart, nor bruise it with his horsemen. This also cometh forth from the Lord of hosts, which is wonderful in counsel, and excellent in working. Isaiah 28:23-29. God dispenses His gifts as it pleases Him. He bestows one gift upon one, and another gift upon another, but all for the good of the whole body. It is in God's order that some shall be of service in one line of work, and others in other lines—all working under the selfsame Spirit. The recognition of this plan will be a safeguard against emulation, pride, envy, or contempt of one another. It will strengthen unity and mutual love. (Counsels to Parents, Teachers and Students p. 314)

STUDENTS SHOULD HAVE MANUAL LABOR: Students should have manual work to do, and it will not hurt them if in doing this work they become weary. Do you not think Christ became weary?—Indeed he did. Weariness injures no one. It only makes rest sweeter. (General Conference Bulletin, April 14, 1901)

YOUNG MEN SHOULD LEARN HOW TO COOK AND YOUNG WOMEN SHOULD LEARN HOW TO GARDEN: The young men, as well as the young women are to be taught how to cook; and the young women, as well as the young men, are to take a part in outside work. When this is done, there will be found in our schools in America as healthy a class of students as is found in our school in Cooranbong, where there are few of the students whose health has not been improved by correct habits of life. (General Conference Bulletin, April 14, 1901)

WOMEN TO LEARN GARDENING AND BE INVOLVED IN ACTIVE PHYSICAL WORK AS WELL: There are many things which the women students may also engage in, such as cooking, dressmaking, and gardening. Plants and flowers should be cultivated, strawberries should be planted. Thus the women students may be called out of doors to gain healthful exercise, and to be educated in useful labor. Book binding also, and a variety of trades should be taken up. These will not only give exercise to brain, bone, and muscle, but they will also give knowledge of great value. The greatest curse of our world today is idleness. The students coming to our school have had an abundance of amusements, which serve merely to please and gratify self. They are now to be given a different education, that they may be prepared to go forth from the school prepared for any service. (Manuscript Release 8 p. 152)

EDUCATION FOR WOMEN TO BE COMPREHENSIVE AND PRACTICAL TOO: For the lady students there are many employments which should be provided, that they may have a comprehensive and practical education. They should be taught dressmaking and gardening. Flowers should be cultivated and strawberries planted. Thus, while being educated in useful labor, they will have healthful outdoor exercise. (Counsels for the Church p. 209)

MANUAL TRAINING IN MANY PRACTICAL FIELDS TO RECEIVE FAR MORE ATTENTION THAN IT HAS: Manual training is deserving of far more attention than it has received. Schools should be established that, in addition to the highest mental and moral culture, shall provide the best possible facilities for physical development and industrial training. Instruction should be given in agriculture, manufactures,—covering as many as possible of the most useful trades,—also in household economy, healthful cookery, sewing, hygienic dressmaking, the treatment of the sick, and kindred lines. Gardens, workshops, and treatment rooms should be provided, and the work in every line should be under the direction of skilled instructors. (Education p. 218)

IF ACADEMIC AND PRACTICAL TRAINING CANNOT BOTH BE HAD, THE PRACTICAL WOULD UNHESITATINGLY BE CHOSEN: If young persons can have but one set of faculties disciplined, which is most important, the study of the sciences, with the disadvantages to health and morals under which such knowledge is usually obtained, or a thorough training in practical duties, with sound morals and good physical development? In most cases both may be secured if parents will take a little pains; but if both cannot be had, we would unhesitatingly decide in favor of the latter. (Signs of the Times, August 26, 1886)

THE PRACTICAL, RATHER THAN THEORETICAL, MOST ESSENTIAL TO STUDENTS: Instruct the students not to regard as most essential the theoretical part of their education. Let it be more and more deeply impressed upon every student that we should have an intelligent understanding of how to treat the physical system. And there are many who would have greater intelligence in these matters if they would not confine themselves to years of study without a practical experience. The more fully we put ourselves under the direction of God, the greater knowledge we shall receive from God. Let us say to our students: Keep yourselves in connection with the Source of all power. Ye are laborers together with God. He is to be our chief instructor. (Review and Herald, November 11, 1909)

PRACTICAL EDUCATION TAKES PRECEDENCE OVER BOOK KNOWLEDGE: If the youth can have but a one-sided education, which is of the greater consequence, a knowledge of the sciences, with all the disadvantages to health and life, or a knowledge of labor for practical life? We unhesitatingly answer, the latter. If one must be neglected, let it be the study of books. (Counsels to Parents, Teachers and Students p. 289)

— (2): Hard Work Employed as a Safeguard against Sin

IT WAS GOD'S PLAN TO REMOVE EVIL FROM THE WORLD THROUGH HARD WORK: It was God's purpose to remove by toil the evil which man brought into the world by disobedience. By toil the temptations of Satan might be made ineffectual, and the tide of evil be stayed. The Son of God was given to the world, by his death to make atonement for the sins of the world, by his life to teach men how the plans of the enemy were to be thwarted. Taking upon himself the nature of man, Christ entered into the sympathies and interests of his brethren, and by a life of untiring labor taught how men might become laborers together with God in the building up of his kingdom in the world. (Review and Herald, October 24, 1907)

A SHIELD FROM DEGRADATION: Every student should devote a portion of each day to active labor. Thus habits of industry would be formed and a spirit of self-reliance encouraged, while the youth would be shielded from many evil and degrading practices that are so often the result of idleness. (Patriarchs and Prophets p. 601)

THE ABILITY WORK IS A TALENT—AND WE ARE ACCOUNTABLE FOR IT: In the capital of strength a precious talent has been entrusted to men for labor. This is of more value than any bank deposit, and should be more highly prized.... It is a blessing that cannot be purchased with gold or silver, houses or lands; and God requires it to be used wisely. No man has a right to sacrifice this talent to the corroding influence of inaction. All are as accountable for the capital of physical strength as for their capital of means.... (God's Amazing Grace p. 64)

DANGERS OF FAILURE TO LEARN TO WORK: Notwithstanding all that has been said and written regarding the dignity of manual labor, the feeling prevails that it is degrading. Popular opinion has, in many minds, changed the order of things, and men have come to think that it is not fitting for a man who works with his hands to take his place among gentlemen. Men work hard to obtain money; and having gained wealth, they suppose that their money will make their sons gentlemen. But many such fail to train their sons as they themselves were trained, to hard, useful labor. Their sons spend the money earned by the labor of others, without understanding its value. Thus they misuse a talent that the Lord designed should accomplish much good. (Lift Him Up p. 49)

EXAMPLES OF BIBLE CHARACTERS WHO WORKED: Let them [students] read of the sons of the prophets (2 Kings 6:1-7), students at school, who were building a house for themselves, and for whom a miracle was wrought to save from loss the ax that was borrowed. Let them read of Jesus the carpenter, and Paul the tentmaker, who with the toil of the craftsman linked the highest ministry, human and divine. Let them read of the lad whose five loaves were used by the Savior in that wonderful miracle for the feeding of the multitude; of Dorcas the seamstress, called back from death, that she might continue to make garments for the poor; of the wise woman described in the Proverbs, who seeketh wool, and flax, and worketh willingly with her hands; who looketh well to the ways of her household, and eateth not the bread of idleness (Proverbs 31:13, 27) (Education pp. 215-217)

GOD PLANNED FOR ADAM AND EVE TO BE ACTIVE AND USEFUL: In creating Adam and Eve, God designed that they should be active and useful. The holy pair was placed in Paradise and surrounded with everything that was pleasant to the eye or good for food. A beautiful garden was planted for them in Eden. In it were stately trees of every description, all that could serve for use or ornament. Flowers of rare loveliness, and of every tint and hue, perfumed the air. Merry songsters of varied plumage caroled joyous songs of praise of the Creator. Paradise delighted the senses of the holy pair, but this was not enough; they must have something to call into play the wonderful human organism. He who formed them knew what would be for their good; and had happiness consisted in doing nothing, they, in their state of holy innocence, would have been left unemployed. But no sooner were our first parents created than God appointed them their work. They were to find employment and happiness in tending the things which God had created, and their wants were to be abundantly supplied from the fruits of the garden. (Christ Triumphant p. 20)

MAN WAS NEVER MEANT TO BE IDLE: The Lord's purposes are not the purposes of men. He did not design that men should live in idleness. In the beginning He created man a gentleman; but though rich in all that the Owner of the universe could supply, Adam was not to be idle. No sooner was he created than his work was given him. He was to find employment and happiness in tending the things that God had created, and in response to his labor his wants were to be abundantly supplied from the fruits of the Garden of Eden. (Lift Him Up p. 49)

PLEASANT WORK IN THE GARDEN OF EDEN ASSIGNED FROM THE VERY START: Although everything God had made was in the perfection of beauty, and there seemed nothing wanting upon the earth which God had created to make Adam and Eve happy, yet He manifested His great love to them by planting a garden especially for them. A portion of their time was to be occupied in the happy employment of dressing the garden, and a portion in receiving the visits of angels, listening to their instruction, and in happy meditation. Their labor was not wearisome but pleasant and invigorating. This beautiful garden was to be their home. In this garden the Lord placed trees of every variety for usefulness and beauty. There were trees laden with luxuriant fruit, of rich fragrance, beautiful to the eye, and pleasant to the taste, designed of God to be food for the holy pair. There were the lovely vines which grew upright, laden with their burden of fruit, unlike anything man has seen since the fall. The fruit was very large and of different colors; some nearly black, some purple, red, pink, and light green. This beautiful and luxuriant growth of fruit upon the branches of the vine was called grapes. They did not trail upon the ground, although not supported by trellises, but the weight of the fruit bowed them down. It was the happy labor of Adam and Eve to form beautiful bowers from the branches of the vine and train them, forming dwellings of nature's beautiful, living trees and foliage, laden with fragrant fruit. It was the design of God that man should find happiness in the employment of tending the things He had created, and that his wants should be met with the fruits of the trees of the garden.... (Conflict and Courage p. 12)

WORK GIVEN AS A BLESSING IN EDEN TO BUILD BODY, CHARACTER, AND MIND: To Adam and Eve was committed the care of the garden, to dress it and to keep it (Genesis 2:15). Though rich in all that the Owner of the universe could supply, they were not to be idle. Useful occupation was appointed them as a blessing, to strengthen the body, to expand the mind, and to develop the character. (To Be Like Jesus p. 254)

IDLENESS—NOT LABOR—DEGRADES: Notwithstanding all that has been said and written concerning the dignity of labor, the feeling prevails that it is degrading. Young men are anxious to become teachers, clerks, merchants, physicians, lawyers, or to occupy some other position that does not require physical toil. Young women shun housework and seek an education in other lines. These need to learn that no man or woman is degraded by honest toil. That which degrades is idleness and selfish dependence. Idleness fosters self-indulgence, and the result is a life empty and barren—a field inviting the growth of every evil.... (To Be Like Jesus p. 234)

THE BEST RECREATION FOUND IN TASKS THAT ARE USEFUL: As a rule, the exercise most beneficial to the youth will be found in useful employment. The little children find both diversion and development in play; and their sports should be such as to promote not only physical, but mental and spiritual growth. As they gain strength and intelligence, the best recreation will be found in some line of effort that is useful. That which trains the hand to helpfulness, and teaches the young to bear their share of life's burdens, is most effective in promoting the growth of mind and character. (Education p. 214-215)

USEFUL OCCUPATION IS ONE OF THE GREATEST SAFEGUARDS FOR THE YOUTH AGAINS DAYDREAMING, BAD HABITS, AND SIN: One of the surest safeguards for the young is useful occupation. Had they been trained to industrious habits, so that all their hours were usefully employed, they would have no time for repining at their lot or for idle daydreaming. They would be in little danger of forming vicious habits or associations. Let the youth be taught from childhood that there is no excellence without great labor. Aspirations for eminence will not avail. Young friends, the mountain-top cannot be reached by standing still, and wishing yourselves there. You can gain your object only by taking one step at a time, advancing slowly perhaps, but holding every step gained. It is the energetic, persevering worker that will scale the Alps. Every youth should make the most of his talents, by improving to the utmost present opportunities. He who will do this, may reach almost any height in moral and intellectual attainments. But he must possess a brave and resolute spirit. He will need to close his ears to the voice of pleasure; he must often refuse the solicitations of young companions. He must stand on guard continually, lest he be diverted from his purpose. (Review and Herald, September 13, 1881)

MANUAL OCCUPATION ESSENTIAL: Manual occupation for the youth is essential. The mind is not to be constantly taxed to the neglect of the physical powers. Ministers and teachers need to learn in regard to these things, and they need to practice well. We need a change of employment, and nature is a living, healthful teacher. (Fundamentals of Christian Education p. 321)

GARDENING—AND KEEPING THE SYSTEM IN ORDER—IS DOING THE WORK OF GOD: Brethren, when you take time to cultivate your garden, thus gaining the exercise needed to keep the system in good working order, you are just as much doing the work of God as in holding meetings. (Gospel Workers p. 240)

THE SENTENCE OF HARD WORK METED OUT AFTER SIN: And unto Adam he said, Because thou hast hearkened unto the voice of thy wife, and hast eaten of the tree, of which I commanded thee, saying, Thou shalt not eat of it: cursed is the ground for thy sake; in sorrow shalt thou eat of it all the days of thy life; Thorns also and thistles shall it bring forth to thee; and thou shalt eat the herb of the field; In the sweat of thy face shalt thou eat bread, till thou return

Chapter #5: Characteristics of Model Schools that Follow God's Plan

unto the ground; for out of it wast thou taken: for dust thou art, and unto dust shalt thou return. (Genesis 3:17-19)

THE THORN AND THISTLE WERE GIVEN WITH MAN'S GOOD IN MIND: It is written that God cursed the ground for man's sake. Genesis 3:17. The thorn and the thistle—the difficulties and trials that make his life one of toil and care—were appointed for his good as a part of the training needful in God's plan for his uplifting from the ruin and degradation that sin has wrought. (Steps to Christ p. 9)

IMPOSSIBLE TO DEGRADE SUCH A WOMAN: Girls should be taught to work, to study to some purpose, to live for some object, to trust in God and fear Him, and to respect their parents. Then as they advance in years, they will grow more pure-minded, self-reliant, and beloved. It will be impossible to degrade such a woman. She will escape the temptations and trials that have been the ruin of so many. (Child Guidance p. 140)

HARD WORK HELPS SAFEGUARD AGAINST IMPURE THOUGHTS AND ACTIONS, AS WELL AS OVERTAXING THE BRAIN: The proportionate taxation of the powers of mind and body will prevent the tendency to impure thoughts and actions. Teachers should understand this. They should teach students that pure thoughts and actions are dependent on the way in which they conduct their studies. Conscientious actions are dependent on conscientious thinking. Exercise in agricultural pursuits and in the various branches of labor is a wonderful safeguard against undue brain taxation. No man, woman, or child who fails to use all the powers God has given him can retain his health. He cannot conscientiously keep the commandments of God. He cannot love God supremely and his neighbor as himself. . . . (Manuscript Release 11 p. 174)

JESUS, OUR EXAMPLE, ENGAGED IN HUMBLE PHYSICAL LABOR: Health and a clear conscience will attend those who work faithfully, keeping the glory of God in view. There are many who are mere fragments of men. In Christ is seen the perfection of Christian character. He is our pattern. His life was not a life of indolence or ease. He lived not to please Himself. He was the Son of the infinite God, yet He worked at the carpenter's trade, with His father. As a member of the home firm, He faithfully acted His part in helping to support the family. . . . (Manuscript Release 11 p. 174)

MANUAL LABOR ESSENTIAL TO THE EDUCATION MOST FAVORABLE TO SPIRITUAL ADVANCEMENT: It reveals cowardice to move so slowly and uncertainly in the labor line—that line which will give the very best kind of education. Look at nature. There is room within her vast boundaries for schools to be established where grounds can be cleared and land cultivated. This work is essential to the education most favorable to spiritual advancement; for nature's voice is the voice of Christ, teaching us innumerable lessons of love and power and submission and perseverance. (Testimonies to the Church Vol. 6 p. 178)

THE HOPE OF ADVANCING GOD'S CAUSE IS WRAPPED UP IN CREATING A NEW MORAL TASTE FOR WORK: The hope of advancing the cause of God in this country is in creating a new moral taste in love of work, which will transform mind and character. (Special Testimonies on Education p. 99)

TEMPATIONS FOR THE YOUTH ABOUND IN THE CITIES—THE SIMPLICITY AND HARD WORK OF COUNTRY LIFE IS A SAFEGUARD: Many parents remove from their country homes to the city, regarding it as a more desirable or profitable location. But by making this change they expose their children to many and great temptations. The boys have no employment, and they obtain a street education, and go on from one step in depravity to another, until they lose all interest in anything that is good and pure and holy. How much better had the parents remained with their families in the country, where the influences are most favorable for physical and mental strength. Let the youth be taught to labor in tilling the soil, and let them sleep the sweet sleep of weariness and innocence. (Review and Herald, September 13, 1881)

SOME INVENTIONS THAT MAKE FARMING EASIER ALSO LEAD TO FORGETFULNESS OF GOD: The greater the length of time the earth has lain under the curse, the more difficult has it been for man to cultivate it, and make it productive. As the soil has become more barren, and double labor has had to be expended upon it, God has raised up men with inventive faculties to construct implements to lighten labor on the land groaning under the curse. But God has not been in all man's inventions. Satan has controlled the minds of men to a great extent, and has hurried men to new inventions which has led them to forget God. (Spiritual Gifts Vol. 4a p. 155)

Characteristic #4: Rural or Country Location

SCHOOLS SHOULD BE AWAY FROM THE CITIES, WHERE THERE IS LAND FOR CULTIVATION: God bids us establish schools away from the cities, where, without let or

Chapter #5: Characteristics of Model Schools that Follow God's Plan

hindrance, we can carry on the work of education upon plans that are in harmony with the solemn message that is committed to us for the world. Such an education as this can best be worked out where there is land to cultivate, and where the physical exercise taken by the students can be of such a nature as to act a valuable part in their character-building, and to fit them for usefulness in the fields to which they shall go. (Testimonies Relevant to Emmanual Missionary College)

OUR SCHOOLS SHOULD BE OUT IN THE COUNTRY: The light that has been given me is that Battle Creek has not the best influence over the students in our school. There is altogether too congested a state of things. The school, although it will mean a fewer number of students, should be moved out of Battle Creek. Get an extensive tract of land, and there begin the work which I entreated should be commenced before our school was established here,—to get out of the cities, to a place where the students would not see things to remark upon and criticize, where they would not see the wayward course of this one and that one, but would settle down to diligent study. (General Conference Bulletin, April 14, 1901)

WHENEVER POSSIBLE, SCHOOLS SHOULD BE IN PEACEFUL COUNTRY LOCATIONS: When students enter the school to obtain an education, the instructors should endeavor to surround them with objects of the most pleasing, interesting character, that the mind may not be confined to the dead study of books. The school should not be in or near a city, where its extravagance, its wicked pleasures, its wicked customs and practises will require constant work to counteract the prevailing iniquity, that it may not poison the very atmosphere which the students breathe. All schools should be located, so far as possible, where the eye will rest upon the things of nature instead of clusters of houses. The ever-shifting scenery will gratify the taste, and control the imagination. Here is a living teacher, instructing constantly. (Special Testimonies on Education p. 98)

THE BEST LOCATION FOR SCHOOLS: Our schools should be located away from the cities, on a large tract of land, so that the students will have opportunity to do manual work. They should have opportunity to learn lessons from the objects which Christ used in the inculcation of truth. He pointed to the birds, to the flowers, to the sower and the reaper. In schools of this kind not only are the minds of the students benefited, but their physical powers are strengthened. All portions of the body are exercised. The education of mind and body is equalized. The body needs a great deal more care than it gets. There are men here who are suffering, O so much, because they are not faithful stewards of their bodies. God wants you to use every means in your power to care for the wonderful machinery which he has given you. Let no part of it rust from inaction. (General Conference Bulletin, April 14, 1901)

Hope in the Soil

I could not sleep past two o'clock this morning. During the night season I was in council. I was pleading with some families to avail themselves of God's appointed means, and get away from the cities to save their children. (Review and Herald, December 11, 1900)

The time is fast coming when the controlling power of the labor unions will be very oppressive. Again and again the Lord has instructed that our people are to take their families away from the cities, into the country, where they can raise their own provisions; for in the future the problem of buying and selling will be a very serious one. We should now begin to heed the instruction given us over and over again: Get out of the cities into rural districts. (Country Living, pp. 9-10)

The Protestant world have set up an idol Sabbath in the place where God's Sabbath should be, and they are treading in the footsteps the Papacy. For this reason I see the necessity of the people of God moving out of the cities into retired country places, where they may cultivate the land and raise their own produce. (Country Living, p. 21)

Believers who are now living in the cities will have to move to the country, that they may save their children from ruin. Attention must be given to the establishment of industries in which these families can find employment. Those who have charge of the schoolwork at _____ and _____ should see what can be done by these institutions to establish such industries, so that our people desiring to leave the cities, can obtain modest homes without a large outlay of means, and can find employment. (Country Living, p. 19)

Instead of the crowded city seek some retired situation where your children will be, so far as possible, shielded from temptation, and there train and educate them for usefulness. (Signs of the Times p. 232)

A MINISTRY TO THE CITIES FROM OUTPOSTS: Repeatedly the Lord has instructed us that we are to work the cities from outpost centers. In these cities we are to have houses of worship, as memorials for God; but institutions for the publication of our literature, for the healing of the sick, and for the training of workers, are to be established outside the cities. Especially is it important that our youth be shielded from the temptations of city life. (Selected Messages Vol. 2 p. 358)

RETIRED MOUNTAIN PLACES, APART FROM THE WORLD, IDEAL AS LOCATIONS FOR ADVANCED SCHOOLS: The time has come when every advantage to be gained for the furtherance of the work should be recognized, for we need all the strength we can obtain. Christ is soon coming, and Satan knows that his time is short. As we draw near to the close of time the cities will become more and more corrupt, and more and more objectionable as places for establishing centers of our work. The dangers of travel will increase, confusion and drunkenness will abound. If there can be found places in retired mountain regions where it would be difficult for the evils of the cities to enter, let our people secure such places for our sanitariums and advanced schools. The two institutions may be far enough apart so that there need be no confusion. (Manuscript Release 10 p. 260)

JOHN THE BAPTIST WAS EDUCATED IN THE WILDERNESS, AMIDST THE PEACEFUL SCENES OF THE COUNTRY: John, the greatest prophet that has ever been delegated to bear a startling message to the world, obtained his education in the wilderness. The scenery of nature was before him as an open book, and God was his teacher. The flattering temptations that come to those who are crowded in the cities did not reach John in the wilderness. His eyes rested upon scenes that were pure and natural, and revealed the character of God to his soul, so that he looked up from nature to nature's God. (Manuscript Release 13 p. 353)

MORE ON JOHN THE BAPTIST'S TRAINING: His education was gained from heaven-appointed sources, from useful work, from the study of the Scriptures and of nature, and from the experiences of life—God's lesson books, full of instruction to all who bring to them the willing heart, the seeing eye, and the understanding heart. (Education p. 77)

SCHOOLS IN THE COUNTRY MUCH BETTER SUITED FOR CHILDREN: It seems cruel to establish our schools in the cities, where the students are prevented from learning the precious lessons taught by nature. It is a mistake to call families into the city, where children and youth breathe an atmosphere of corruption and crime, sin and violence, intemperance and ungodliness. Oh, it is a terrible mistake to allow children to come in contact with that which makes such a fearful impression on their senses. Children and youth cannot be too fully guarded from familiarity with the pictures of iniquity as common as in all large cities. (Spalding and Magan p. 186)

SCHOOLS TO BE AWAY FROM THE CITIES: God bids us establish schools away from the cities, where, without let or hindrance, we can carry on the education of students upon plans that are in harmony with the solemn message committed to us for the world. Such an education as this can best be worked out where there is land to cultivate and where the physical exercise taken by the students can be of such a nature as to act a valuable part in their character building and fit them for usefulness in the fields to which they shall go. (Counsels to Parents, Teachers and Students p. 532)

SCHOOLS TO BE OUT OF THE CITIES: Be assured that the call is for our people to locate miles away from the large cities. One look at San Francisco as it is today would speak to your intelligent minds, showing you the necessity of getting out of the cities. Do not establish institutions in the cities, but seek a rural location. The call is, Come out from among them, and be ye separate. The very atmosphere of the city is polluted. Let your schools be established away from the cities, where agricultural and other industries can be carried on. (Manuscript Release 21 p. 90)

OUT OF THE CITIES IS THE MESSAGE FOR PARENTS AND SCHOOLS: Let parents understand that the training of their children is an important work in the saving of souls. In country places abundant useful exercise will be found in doing those things that need to be done, and which will give physical health by developing nerve and muscle. Out of the cities is my message for the education of our children. (Manuscript Release 10 p. 260)

STUDENTS TO BE EDUCATED TO EDUCATE THE LAND IN A RURAL SETTING: The good hand of the Lord has been with our people in the selection of a good place to locate the school [in Berrien Springs]. This place corresponds to the representations given me as to where the school would be located. It is away from the cities, and there is an abundance of land for agricultural purposes, and room so that the houses will not need to be built one close to another. There is plenty of ground where students shall be educated to educate the land. Ye and God's husbandry: ye are God's building. (Spalding and Magan p. 203)

THE IMPORTANCE OF SCHOOLS IN THE COUNTRY FOR YOUTH WHO ARE NOT YET SETTLED INTO THE TRUTH: The youth who attend our school for the first time, are not prepared to exert a correct influence in any city as lights shining amid the darkness. They will not be prepared to reflect light until the darkness of their own erroneous education is dispelled. In the future our school will not be the same as it has been in the past. Among the students there have been reliable, experienced men who have taken advantage of the opportunity to gain more knowledge in order to do intelligent work in the cause of God. These have been a help in the school, for they have been as a balance-wheel; but in the future the school will consist mostly of those who need to be transformed in character, and who will need to have much patient labor bestowed upon them; they have to unlearn, and learn again. It will take time to develop the true missionary spirit, and the farther they are removed from the cities and the temptations that are flooding them, the more favorable will it be for them to obtain the true knowledge and develop well-balanced characters. (Special Testimonies on Education p. 102)

GOD WILL HELP WHEN SURROUNDINGS ARE NOT IDEAL: We may be placed in trying positions, for many cannot have their surroundings what they would; and whenever duty calls us, God will enable us to stand uncorrupted, if we watch and pray, trusting in the grace of Christ. But we should not needlessly expose ourselves to influences that are unfavorable to the formation of Christian character. (Patriarchs and Prophets p. 169)

AVOIDING A STREET EDUCATION: It would be well for you to lay by your perplexing cares and find a retreat in the country, where there is not so strong an influence to corrupt the morals of the young. True, you would not be entirely free from annoyances and perplexing cares in the country; but you would there avoid many evils and close the door against a flood of temptations which threaten to overpower the minds of your children. They need employment and variety. The sameness of their home makes them uneasy and restless, and they have fallen into the habit of mingling with the vicious lads of the town, thus obtaining a street education.... (Adventist Home p. 141)

RICH BLESSINGS AWAIT THOSE WHO MOVE INTO THE COUNTRY: Do not consider it a privation when you are called to leave the cities and move out into the country places. Here there await rich blessings for those who will grasp them. (Adventist Home p. 141)

ONE BY ONE THE CITIES WILL BE DESTROYED: The Lord calls for His people to locate away from the cities, for in such an hour as ye think not, fire and brimstone will be rained from heaven upon these cities. Proportionate to their sins will be their visitation. When one city is destroyed, let not our people regard this matter as a light affair, and think that they may, if favorable opportunity offers, build themselves homes in that same destroyed city. (Manuscript Release 21 p. 90)

OUT OF THE CITIES—THE MESSAGE AT THIS TIME: The outlook for establishing a sanitarium at Adelaide is much more favorable than the outlook for establishing one at Melbourne. The city of Melbourne is not the place to establish a sanitarium. It has been plainly presented to me that the sanitarium which you are planning to establish should be located in the most healthful place you can secure. But my warning is that of the angel who, standing in Melbourne, said in a clear, distinct voice, Establish not schools or sanitariums in the cities. In the future, cities will certainly feel the terrible results of earthquakes and fires. Cities will be destroyed by flood and by lightnings. Out of the cities, is my message at this time. (Manuscript Release 21 p. 90)

— (1): Plenty of Land Reserved for Use by the School

ADEQUATE LAND NEEDED TO START A SCHOOL: Our school should be located where the students can receive an education broader than that which the mere study of books will give. They must have such a training as will fit them for acceptable service if they are called to do pioneer work in mission fields either in America or in foreign countries. There must be land enough to give an experience in the cultivation of the soil, and to help largely in making the institution self-supporting. (Manuscript Release 1 p. 333)

ROOM FOR ORCHARDS AND VEGETABLE GARDENS NEEDED: In establishing schools and sanitariums, enough land should be purchased to provide for the carrying out of the plans that the Lord has outlined for these institutions. Provision should be made for the raising of fruit and vegetables, and, wherever possible, sufficient land should be secured so that others may not erect, near the institution, buildings of an objectionable character. (Gospel Workers p. 457)

EDUCATION WOULD BE GREATLY AIDED IF ALL SCHOOLS INCLUDED LAND FOR CULTIVATION: With the question of recreation the surroundings of the home and the school have much to do. In the choice of a home or the location of a school these things should be considered. Those with whom mental and physical well-being is of greater moment than money or the claims and customs of society, should seek for their children the benefit of nature's teaching, and recreation amidst her surroundings. It would be a great aid in educational work could every school be so situated as to afford the pupils land for cultivation, and access to the fields and woods. (Education p. 211)

GOD'S SCHOOLS NEED ELBOW ROOM: We must have room to keep ourselves distinct as a Sabbath-keeping people. The Lord has given directions that we are to make provision which will prevent our being harassed and inconvenienced by having to crowd in with unbelievers. I wish I might make on your minds the impression that has been made on mine regarding this matter. (The Purchase of Land at Loma Linda p. 19)

EVERY SCHOOL SHOULD HAVE A FARM CONNECTED WHICH IS REGARDED AS GOD'S OWN SCHOOLROOM, A LESSONBOOK FROM WHICH TO STUDY: The youth should be instructed in a similar way. They should be taught to till the soil. It would be well if there were, connected with every school, lands for cultivation. Such lands should be regarded as God's own schoolroom. The things of nature should be looked upon as a lesson book which

His children are to study, and from which they may obtain knowledge as to the culture of the soul. (Christ's Object Lessons p. 87)

GOD WANTS US TO HAVE PEACE, GLADNESS, AND ELBOW ROOM: He [God] wants us to live where we can have elbow room. His people are not to crowd into the cities. He wants them to take their families out of the cities, that they may better prepare for eternal life. In a little while they will have to leave the cities. These cities are filled with wickedness of every kind—with strikes and murders and suicides. Satan is in them, controlling men in their work of destruction. Under his influence they kill for the sake of killing, and this they will do more and more. Every mind is controlled either by the power of Satan or the power of God. If God controls our minds, what shall we be?—Christian gentlemen and Christian ladies. God can fill our lives with His peace and gladness and joy. He wants His joy to be in us, that our joy may be full. (Manuscript Release 12 p. 30)

GOD'S DIRECTION NEEDED IN ALL AREAS OF LIFE—INCLUIDNG LOCATING LAND: The Lord knows what is best for His work. That which was, as it were, a hiding place in the wilderness has proved to be a profitable tract of land. And we have learned that if we would have a rich experience in our Christian life, we must let the Lord direct. (Manuscript Release 15 p. 56)

HAD SCHOOLS BEEN ESTABLISHED ON LARGE TRACTS OF LAND, THE A DIFFERENT ORDER OF THINGS WOULD NOW EXIST: Years ago schools should have been established on large tracts of land, where children could have been educated largely from the book of nature. Had this been done, what a different condition of things there would now be in our churches. We are in need of being uplifted, cleansed, purified. In our conversation we are altogether too cheap and common. There are tares growing among the wheat, and too often the tares overtop the wheat. (Spalding and Magan p. 191)

HAPPINESS AT THE DECISION TO PLACE A SCHOOL IN A FARMING DISTRICT: I rejoiced when I heard that the Battle Creek School was to be established in a farming district. I know that there will be less temptation there for the students than there would be in the cities that are fast becoming as Sodom and Gomorrah, preparing for destruction by fire. The popular sentiment is that cities should be chosen as locations for our schools. But God desires us to leave the sin-polluted atmosphere of the cities. It is his design that our schools shall be established where the atmosphere is purer. (Spalding and Magan p. 191)

HOUSES NOT TO BE NEAR THE SCHOOL: Plans were laid to build cottages on the school campus [Avondale College]. I was glad I was here at the time that this subject was brought up, for I had something to say. I told them that the grounds were not to be occupied by buildings. The land is to be our lesson book. After being cleared, it is to be cultivated. Orange, lemon, peach, apricot, nectarine, plum, and apple trees are to occupy the land, with vegetable gardens, flower gardens, and ornamental trees. Thus this place is to be brought as near as possible to the presentation that passed before me several times, as the symbol of what our school and premises should be. Dwelling houses, fenced allotments for families were not to be near our school buildings. This place must by the appointment of God be a representation of what school premises should be—a delight to the eyes. (Manuscript Release 11 p. 185)

HOUSING FOR BELIEVERS NOT TO BE TOO NEAR SCHOOL PROPERTY: In planning for the erection of cottages for our brethren and sisters who may move there, be careful not to allow buildings to be put up too near the school property. Try to secure the land lying near the school, so that it will be impossible for houses to be built close to the campus. The land may be used for agricultural purposes. Later on, you may find it advisable to introduce various trades for the employment and training of the students; but at present about all that you can do is to teach them how to cultivate the land, so that it shall yield its fruit. (Manuscript 54, 1903, pp. 1-4 "The Work of Our Fernando School")

IF EXPENSES MUST BE CUT AT THE ESTABLISHMENT OF A SCHOOL, SPEND MORE ON LAND AND LESS ON BUILDINGS: In establishing schools, enough land should be secured to give the students opportunity to gain a knowledge of agriculture. If it is necessary to curtail the expense anywhere, let it be on the buildings. There should be no failure to secure land; for from the cultivation of the soil, the students are to learn lessons illustrating the truths of the word of God, truths that will help them to understand the work of the Creator. (Manuscript Release 6 p. 411)

LAND CLOSE TO THE SCHOOL IN COORANBONG SHOULD BE RETAINED BY THE SCHOOL: Let the lands near the school and the church be retained. Those who come to settle in Cooranbong can, if they choose, find for themselves homes nearby, or on portions of, the Avondale Estate. But the light given to me is that all that section of land from the school orchard to the Maitland road, and extending on both-sides of the road from the meeting house to the school, should become a farm and a park, beautified with fragrant flowers and ornamental trees. There should be fruit orchards and every kind of produce cultivated that is adapted to this soil, that this place may become an object lesson to those living close by and afar off. (<u>Australasian Union Conference Record</u>, July 31, 1899)

LAND NEAR THE SCHOOL TO BE CULTIVATED BY THE STUDENTS AND A LESSONBOOK FOR TEACHERS TO USE: If you should allow the land near the school to be occupied with private houses and then be driven to select for cultivation other land at a distance from the school, it would be a great mistake and one always to be regretted. All the land near the building is to be regarded as the school farm, where the youth can be educated under well-qualified superintendents. The youth who shall attend our schools need all the land nearby. They are to plant it with ornamental and fruit trees, and to cultivate garden produce. The school farm is to be regarded as a lesson book in nature from which the teachers may draw their object lessons. Our students are to be taught that Christ, who created the world and all things that are therein, is the life and light of every living thing. The life of every child and youth who is willing to grasp the opportunities of receiving a proper education will be made thankful and happy while at school by the things upon which his eyes shall rest. (<u>Testimonies to the Church Vol. 6</u> p. 181)

LAND NEAR THE SCHOOL TO BE RESERVED FOR THE BENEFIT OF THE SCHOOL: We have had an experience to teach us what this means. Nearly one year ago, as we were living the last days of the old year, my heart was in a burdened condition. I had matters opening before me in regard to the dangers of disposing of land near the school for dwelling houses. We seemed to be in a council meeting, and there stood One in our midst who was expected to help us out of our difficulties. The words spoken were plain and decided, This land, by the appointment of God, is for the benefit of the school. You have recently had an evidence of human nature, what it will reveal under temptation. The more families you settle about the school buildings, the more difficult it will be for teachers and students. (<u>Manuscript 115</u>, 1898)

LESSON FROM NATURE ABOUT UNCRAMPED ROOTS AND UNCRAMPED MINDS: Well, the school [Avondale College] has made an excellent beginning. The students are learning how to plant trees, strawberries, etc.; how they must keep every sprangle and fiber of the roots uncramped to give them a chance to grow. Is not this a most precious lesson as to how to treat the human mind, and the body as well, not to cramp any of the organs of the body, but give them ample room to do their work? The mind must be called out, its energies taxed. (<u>Manuscript Release 11</u> p. 183)

LAND AROUND THE SCHOOL TO BE RESERVED AS THE LORD'S FARM: On several occasions the light has come to me that the land around our school (Avondale) is to be used as the Lord's farm. In a special sense portions of this farm should be highly cultivated. Spread out before me I saw the land planted with every kind of fruit tree that will bear fruit in this locality; there were also vegetable gardens, where seeds were sown and cultivated. (Testimonies to the Church Vol. 6 p. 185)

ORCHARDS AND EVERY TYPE OF PRODUCE THAT CAN BE CULTIVATED, SHOULD BE: There should be orchards, and every kind of produce should be cultivated that is adapted to the soil, that this place may become an object lesson to those living close by and far off. (Testimonies to the Church Vol. 6 p. 187)

LOMA LINDA HARDLY COMPLETE WITHOUT CONTROL OF A NEARBY ORCHARD: Recently the question arose about securing more of the nearby land that is for sale. One piece, a tract of eighty-six acres, has already been purchased, and there is now on sale another tract of forty-seven acres joining the Loma Linda property. Because this piece of land is so near to our Loma Linda buildings, we do not want to see it sold to unbelievers, who will divide it up, and sell it to those who may desire to crowd into this neighborhood. In the night season I was talking to our brethren, telling them that this must not be allowed, and pointing out what unfavorable results would follow. If this piece of land should be purchased by unbelievers, and divided up and sold to those who would be no help to our work, the injury to Loma Linda would be serious and lasting. I cannot bear the thought of this. Cannot a group of individuals who are alive to the vital interests of the Lord's work, unite together and make this land our property? Then if we wish to sell a portion of it, let it be sold to our people. There is an orange orchard on the place, and this could be handled to advantage by the Sanitarium. The institution is hardly complete without the control of this orchard. (The Purchase of Land at Loma Linda p. 18)

PROPER USE OF SCHOOL LAND: Last Friday night after retiring, a great burden came upon me. I could not sleep until midnight. About the time of the beginning of the Sabbath, I lay down upon the lounge, and (an unusual thing for me to do) fell asleep. Then some things were presented before me. Some persons were selecting allotments of land, on which they purposed to build their homes, and One stood in our midst and said, You are making a great mistake which you will have cause to regret. This land is not to be occupied with buildings except to provide the facilities essential for the teachers and students of the school. This is the school farm. This land is to be reserved as an acted parable to the students. They are not to look upon the school land as a common thing, but as a lesson book which the Lord would have them study. Its lessons will impart knowledge in the spiritual culture of the soul. For you to settle this land with private houses, and then be driven to select other land at a distance for school purposes would be a great mistake, always to be regretted. All the land upon the ground that is not needed for buildings is to be considered the school farm, where youth may be educated under well-qualified superintendents.. . . . The Lord would have the school grounds dedicated to Him as His own school room. The church premises are not to be invaded with houses. We are located where there is plenty of land. . . . (Manuscript Release 8 p. 264)

SCHOOLS SHOULD HAVE PLENTY OF LAND AROUND THEM AND PLACES WHERE STUDENTS CAN WORK, LEARN FRUGALITY AND HELP THEMSELVES: In connection with our schools, ample grounds should be provided. There are some students who have never learned to economize, and have always spent every shilling they could get. These should not be cut off from the means of gaining an education. Employment should be furnished them, and with their study of books should be mingled a training in industrious, frugal habits. Let them learn to appreciate the necessity of helping themselves. (Special Testimonies on Education p. 45)

THE SELLING OF LANDS CLOSE TO THE SCHOOL, AND FAILURE TO FURNISH STUDENTS WITH OUTDOOR EMPLOYMENT LED TO SLOW AND IMPERFECT ADVANCEMENT IN SPIRITUAL THINGS: It is God's desire that greater attention shall be paid to the spiritual necessities of the children and youth in the Healdsburg school, and in all our schools. When the managers of our schools make up their minds to carry out the principles which for years God has been presenting to them, they will be far better prepared to give attention to the spiritual needs of the students. If in the past, those in charge of the Healdsburg school had had spiritual foresight, they would have secured the land near the school home, which is now occupied by houses. The failure to furnish the students with outdoor employment, in the cultivation of the soil, is making their advancement in spirituality very slow and imperfect. The result of this neglect should lead the teachers to be wise unto salvation. It is a mistake for so many dwelling houses to be crowded close to the school home. This is working greatly to the disadvantage of the students. A lack of wisdom was shown by the failure to secure the land round the school home. This will make the work of preserving order and maintaining discipline harder than it otherwise would be. But order must be preserved at any cost, and the workers in the school must plan how this shall be done most successfully. (Manuscript 11, 1901, pp. 6-7)

ABUNDANCE OF LAND AMONG FEATURES NEEDED FOR A SCHOOL'S LOCATION: Consideration of what I saw, and the description given of other parts of the property made it plain that here were many most precious advantages. It was away from the strong temptations of city life. There was abundance of land for cultivation, and the water

Hope in the Soil

advantages were very valuable. All through the mountains there were little valleys where families might locate and have a few acres of land for garden or orchard. The many pipes laid over the grounds made it possible to use water freely both for the buildings and for the land. (Manuscript Release 1 p. 334)

Characteristic #5: Every Student Learning a Practical Trade

STUDENTS TO BE TAUGHT THE INDUSTRIAL ARTS: All kinds of industrial employment are to be found for the student. The students are constantly to learn how to use brain, bone, and muscle, taxing all harmoniously and equally. (Letter 84, 1898, to J. H. Kellogg)

PRACTICAL TRADES TO BE TAUGHT IN OUR SCHOOLS: We need more teachers and more talent, to educate the students in various lines, that there may go forth from this place many persons willing and able to carry the knowledge which they have received to others. Lads are to come in from different localities, and nearly all will take the industrial course. This course should include the keeping of accounts, carpenter's work, and everything that is comprehended in farming. Preparation should also be made for the teaching of blacksmithing, painting, shoemaking, cooking, baking, washing, mending, typewriting, and printing. Every power at our command is to be brought into this training work, that students may go forth equipped for the duties of practical life. (Advocate, February 1, 1899)

STUDENTS SHOULD HELP WITH CONSTRUCTION: Cottages and buildings essential to the school work are to be erected by the students themselves. These buildings should not be crowded close together, or located near the school buildings proper. In the management of this work, small companies should be formed who should be taught to carry a full sense of their responsibility. All these things cannot be accomplished at once, but we are to begin to work in faith. (Advocate, February 1, 1899)

YOUNG LADIES ARE NOT EXEMPT: Small fruits should be planted, and vegetables and flowers cultivated, and this work the lady students may be called out of doors to do. Thus, while exercising brain, bone, and muscle, they will also be gaining a knowledge of practical life. (Testimonies to the Church Vol. 6 p. 176)

EVERY STUDENT SHOULD DEVELOP STRONG PROFICIENCY IN AT LEAST ONE TRADE BEFORE LEAVING SCHOOL: The work should have a definite aim and should be thorough. While every person needs some knowledge of different handicrafts, it is indispensable that he become proficient in at least one. Every youth, on leaving school, should have acquired a knowledge of some trade or occupation by which, if need be, he may earn a livelihood. (Education p. 218)

BOOK KNOWLEDGE WITHOUT PRACTICAL TRAINING LEADS TO SUPERFICIAL THINKING: The benefit of manual training is needed also by professional men. A man may have a brilliant mind; he may be quick to catch ideas; his knowledge and skill may secure for him admission to his chosen calling; yet he may still be far from possessing a fitness for its duties. An education derived chiefly from books leads to superficial thinking. Practical work encourages close observation and independent thought. Rightly performed, it tends to develop that practical wisdom which we call common sense. It develops ability to plan and execute, strengthens courage and perseverance, and calls for the exercise of tact and skill. (Education p. 220)

STUDENTS TO LEARN HOW TO WORK WITH THEIR HANDS TO EARN THEIR DAILY BREAD: The students are to be given a training in those lines of work that will help them to be successful laborers for Christ. They are to be taught to be separate from the customs and practices of the world. They are to be taught how to present the truth for this time, and how to work with their hands and with their heads to win their daily bread, that they may go forth to teach their own people. The bread-winning part of the work is of the utmost importance. They are to be taught also to appreciate the school as a place in which they are given opportunity to obtain a training for service. (Manuscript Release 2 p. 68)

PRACTICAL SKILLS TO INCLUDE EXPERIENCE IN CANNING AND DRYING FRUITS: Wherever fruit can be grown in abundance, a liberal supply should be prepared for winter, by canning or drying. (Testimony Studies on Diets and Foods p. 82)

BOYS SHOULD ALSO LEARN ABOUT HOUSEHOLD DUTIES: Since both men and women have a part in homemaking, boys as well as girls should gain a knowledge of household duties.... Let the children and youth learn from the Bible how God has honored the work of the everyday toiler. (To Be Like Jesus p. 234)

COOKING AND USEFUL TRADES TO BE TAUGHT IN OUR SCHOOLS: There should have been experienced teachers to give lessons to young ladies in the cooking department. Young girls should have been taught how to cut, make, and mend garments, and thus become educated for the practical duties of life. For young men, there should have been establishments where they could learn different trades, which would bring into exercise their muscles as well as their mental powers. (Counsels to Parents, Teachers, and Students p. 289)

VARIOUS INDUSTRIES TO BE TAUGHT IN OUR SCHOOLS: Various industries should be carried on in our schools. The industrial instruction given should include the keeping of accounts, carpentry, and all that is comprehended in farming. Preparation should be made for the teaching of blacksmithing, painting, shoemaking, and for cooking, baking, washing, mending, typewriting, and printing. Every power at our command is to be brought into this training work, that students may go forth well equipped for the duties of practical life. (Counsels to Parents, Teachers and Students p. 310)

THOROUGH, SYSTEMATIC TRAINING IN MANUAL TRADES TO BE TAUGHT TO YOUNG PEOPLE: A much larger number of young people need to have the advantages of our schools. They need the manual training course, which will teach them how to live an active, energetic life. Under wise, judicious, God-fearing directors, the students are to be taught different kinds of labor. Every branch of the work is to be conducted in the most thorough, systematic way that long experience and wisdom can enable us to plan and execute. (Counsels to Parents, Teachers and Students p. 315)

AGRICULTURE THE MOST VALUABLE LINE OF MANUAL TRAINING: No line of manual training is of more value than agriculture. A greater effort should be made to create and to encourage an interest in agricultural pursuits. Let the teacher call attention to what the Bible says about agriculture: that it was God's plan for man to till the earth; that the first man, the ruler of the whole world, was given a garden to cultivate; and that many of the world's greatest men, its real nobility, have been tillers of the soil. Show the opportunities in such a life. The wise man says, The king himself is served by the field. Ecclesiastes 5:9. Of him who cultivates the soil the Bible declares, His God doth instruct him to discretion, and doth teach him. Isaiah 28:26. And again, Whoso keepeth the fig tree shall eat the fruit thereof. Proverbs 27:18. He who earns his livelihood by agriculture escapes many temptations and enjoys unnumbered privileges and blessings denied to those whose work lies in the great cities. And in these days of mammoth trusts and business competition, there are few who enjoy so real an independence and so great certainty of fair return for their labor as does the tiller of the soil. (Education p. 219)

Chapter #5: Characteristics of Model Schools that Follow God's Plan

INSTRUCTION IN PRACTICAL GARDENING SKILLS:

- Students should be taught how to plant. (Counsels to Parents, Teachers and Students p. 310)
- The special needs of every variety of plant must be studied. Different varieties require different soil and cultivation, and compliance with the laws governing each is the condition of success. (Education pp. 111-112)
- Teach them how to plant and care for orchards. (Ministry of Healing p. 193)
- Teach the people how to cultivate the soil, that it may yield rich treasures. Who will be missionaries to do this work, to teach proper methods to the youth? (Fundamentals of Christian Education p. 324)
- Students should be taught how to gather the harvest. (Counsels to Parents, Teachers and Students p. 310)

Characteristic #6: Farming Supercedes Love of Sports and Amusements

FAITHFUL CARE OVER BUILDINGS AND CULTIVATION OF THE FARM TO NATURALLY SUPERCEDE THE LOVE OF SPORTS AND AMUSEMENT: Let the students who are engaged in building do their tasks with thoroughness, and let them learn from these tasks lessons that will help in their character building. In order to have perfect characters, they must make their work as perfect as possible. Into every line of labor let there be brought that stability which means true economy. If in our schools the land were more faithfully cultivated, the buildings more disinterestedly cared for by the students, the love of sports and amusements, which causes so much perplexity in our schoolwork, would pass away. (Counsels to Parents, Teachers and Students p. 312)

A REPLACEMENT FOR SPORTS AND AMUSEMENTS: In his teachings the Savior represented the world as a vineyard. We would do well to study the parables in which this figure is used. If in our schools the land were more faithfully cultivated, the buildings more disinterestedly cared for by the students, the love of sports and amusements, which causes so much perplexity in our school work, would pass away. (Review and Herald, October 24, 1907)

NO NEED FOR GYMNASTIC EXERCISES: Those who combine useful labor with study have no need of gymnastic exercises. And work performed in the open air is tenfold more beneficial to health than indoor labor. Nothing short of nature's invigorating air and sunshine will fully meet the demands of the system. The tiller of the soil finds in his labor all the movements that were ever practiced in the gymnasium. (Fundamentals of Christian Education p. 73)

AGRICULTURE THE MOST VALUABLE FORM OF RELAXATION FROM STUDY: As a relaxation from study, occupations pursued in the open air, and affording exercise for the whole body, are the most beneficial. (Education p. 219)

SCHOOLS SET FAR AWAY FROM AMUSEMENTS HAVE A BETTER CHANCE OF SUCCESS: For the benefit of our school we knew that we must get away from the cities, where there are so many holidays, and where the interest taken in ball playing, horse racing, and games of every kind, amounts almost to a craze. In the woods we are just where we should be. Not that we expect to get away from Satan and from temptation, but we do hope to be able to teach the youth that there is something satisfying besides amusement. (Manuscript Release 7 p. 254)

TIME THAT WOULD HAVE BEEN SPENT ON SPORTS USED TO DEVELOP A PRACTICAL SKILL: In our school in Australia we educated the youth along these lines, showing them that in order to have an education that is complete, they must divide their time between the gaining of book knowledge and the securing of a knowledge of practical work. Part of each day was spent in manual labor. Thus the students learned how to clear the land, to cultivate the soil, and to build houses; and these lines of work were largely carried on in time that would otherwise have been spent in playing games and seeking for amusement. The Lord blessed the students who devoted their hours to learning lessons of usefulness. (Counsels to Parents, Teachers and Students p. 310)

DAYS OF LEISURE LEAD TO DEPRAVITY, CRUELTY, VIOLENCE, AND CRIME. Through the observance of holidays the people both of the world and of the churches have been educated to believe that these lazy days are essential to health and happiness; but the results reveal that they are full of evil, which is ruining the country. The youth generally are not educated to diligent habits. Cities and even country towns are becoming like Sodom and Gomorrah, and like the world in the days of Noah. The training of the youth in those days was after the same order as children are being educated and trained in this age, to love excitement, to glorify themselves, to follow the imagination of their own evil hearts. Now as then, depravity, cruelty, violence, and crime are the result. (Special Testimonies on Education p. 92)

AMUSEMENTS OF THE WORLD PROMOTE A DISLIKE FOR USEFUL OUTDOOR LABOR: In every school Satan will try to make himself the guide of the teachers who are instructing the students. It is he who would introduce the idea that selfish amusements are a necessity. It is he who would lead students, sent to our schools for the purpose of receiving an education and training for the work of evangelists, ministers, and missionaries, to believe that amusements are essential to keep them in physical health, when the Lord has presented to them that the better way is for them to embrace manual labor in their education, and thus let useful employment take the place of selfish amusements. These amusements, if followed, soon develop a dislike for useful, healthful exercise of body and mind, such as would make students efficient to serve themselves and others. (Manuscript Release 8 p. 151)

EVILS OF CITY AMUSEMENTS: It is Satan's purpose to attract men and women to the cities, and to gain this object he invents every kind of novelty and amusement, every kind of excitement. And the cities of the earth today are becoming as were the cities before the Flood. (Manuscript Release 10 p. 261)

SATAN PLAYING THE GAME OF LIFE AGAINST SOULS: While the youth are becoming expert in games that are of no real value to themselves or to others, Satan is playing the game of life for their souls, taking from them the talents that God has given them; and placing in their stead his own evil attributes. (Counsels to Parents, Teachers and Students pp. 274-275)

PLAY NOT ESSENTIAL: Diligent study is essential, so also is diligent hard work. Play is not essential. Devotion of the physical powers to amusement is not most favorable to a well-balanced mind. If the time employed in physical exercise which step by step leads on to excess were used in working in Christ's lines, the blessing of God would rest upon the worker. (Counsels to Parents, Teachers and Students pp. 308-309)

CHRIST NOT ENGAGED IN AMUSEMENT: I cannot find an instance in the life of Christ where He devoted time to play and amusement. He was the great educator for the present and the future life, yet I have not been able to find one instance where He taught the disciples to engage in amusement in order to gain physical exercise. (Counsels to Parents, Teachers and Students p. 309)

MUCH WILL BE GAINED BY AN EDUCATIONAL EMPHASIS ON PHYSICAL LABOR—NOT SPORTS. STUDENTS WOULD DEVELOP COURAGE, PERSEVERANCE, SELF-RELIANCE, AND MORAL STAMINA. Young persons are naturally active, and if they find no legitimate scope for their pent-up energies after the confinement of the schoolroom, they become restless and impatient of control; they are thus led to engage in the rude, unmanly sports that disgrace so many schools and colleges, and even to plunge into scenes of dissipation. And many who leave their homes innocent, are corrupted by their associations at school. Much could be done to obviate these evils, if every institution of learning would make provision for manual labor on the part of the students,—for actual practice in agriculture and the mechanic arts. Competent teachers should be provided to instruct the youth in various industrial pursuits, as well as in their studies in the school room. While a part of each day is devoted to mental improvement and physical labor, devotional exercises and the study of the Scriptures should not be overlooked. Students trained in this manner would have habits of self-reliance, firmness, and perseverance, and would be prepared to engage successfully in the practical duties of life. They would have courage and determination to surmount obstacles, and moral stamina to resist evil influences. (Signs of the Times, August 26, 1886)

THE EXAMPLE OF JESUS CHRIST: I cannot find an instance in the life of Christ where He devoted time to play and amusement. He was the great educator for the present and the future life, yet I have not been able to find one instance where He taught the disciples to engage in amusement in order to gain physical exercise. (Counsels to Parents, Teachers and Students p. 309)

GOD WILL BLESS STUDENTS WHO DEVOTE TIME TO USEFULNESS: In our school in Australia we educated the youth along these lines, showing them that in order to have an education that is complete, they must divide their time between the gaining of book knowledge and the securing of a knowledge of practical work. Part of each day was spent in manual labor. Thus the students learned how to clear the land, to cultivate the soil, and to build houses; and these lines of work were largely carried on in time that would otherwise have been spent in playing games and seeking for amusement. The Lord blessed the students who devoted their hours to learning lessons of usefulness. (Counsels to Parents, Teachers and Students p. 310)

Chapter #5: Characteristics of Model Schools that Follow God's Plan

Characteristic #7: Teachers and Ministers Working Alongside Students

CHRIST IS TO BE THE REAL TEACHER IN OUR SCHOOLS, WORKING WITH HUMAN TEACHERS AND STUDENTS TO CARRY OUT THE PLAN OF REDEMPTION: If those who have received instruction concerning God's plan for the education of the youth in these last days, will surrender their wills to God, he will teach them his will and his way. Christ is to be the teacher in all our schools. If teachers and students will give him his rightful place, he will work through them to carry out the plan of redemption. (Review and Herald, October 24, 1907)

TEACHERS WORKING ALONGSIDE THE STUDENTS: Let the teachers share the work with the students, and show what results can be achieved through skillful, intelligent effort. (Education p. 219)

TEACHERS NEED TO EDUCATE FAR MORE FROM NATURE THAN THEY DO: Those who educate the youth in book knowledge need physical exercise to strengthen the muscles as much as do our students. Our teachers need to educate far more from nature than they do. Nature is God's great school, and on these grounds resources are found for acquiring greater knowledge of the wonderful works of God. (Manuscript Release 13 p. 352)

OUR EXAMPLE—THE GREATEST TEACHER EVER—DREW LESSONS FROM NATURE: Jesus, the greatest teacher the world ever knew, drew the most valuable illustrations of truth from scenes in nature. Parents, imitate His example, and use the things that delight the senses to impress important truths upon the minds of your children. Take them out in the morning, and let them hear the birds caroling forth their songs of praise. Teach them that we too should return thanks to the bountiful Giver of all for the blessings we daily receive. Teach them that it is not dress that makes the man or the woman, but that it is true goodness of heart. (The Review and Herald, October 27, 1885)

TEACHER'S EXAMPLE NEEDED TO INSPIRE THE STUDENTS: Outdoor exercise, especially in useful labor, is one of the best means of recreation for body and mind; and the teacher's example will inspire his pupils with interest in and respect for manual labor. (Education p. 278)

ENTHUSIASM IMPORTANT: An important element in educational work is enthusiasm. (Education p. 233)

ENTHUSIASM CONTAGIOUS: His (Nehemiah's) hope, his energy, his enthusiasm, his determination, were contagious, inspiring others with the same high courage and lofty purpose. (Prophets and Kings p. 638)

TEACHERS NEED TO WAKE UP TO THE IMPORTANCE OF AGRICULTURE IN THE SCHOOLS: Let the teachers wake up to the importance of this subject, and teach agriculture and other industries that it is essential for the students to understand. Seek in every department of labor to reach the very best results. Let the science of the word of God be brought into the work, that the students may understand correct principles, and may reach the highest possible standard. Exert your God-given abilities, and bring all your energies into the development of the Lord's farm. Study and labor, that the best results and the greatest returns may come from the seed-sowing, that there may be an abundant supply of food, both temporal and spiritual, for the increased number of students that shall be gathered in to be trained as Christian workers. (Testimonies to the Church Vol. 6 p. 191)

EDUCATE YOUTH TO SEE THE HOE AND SHOVEL AS INSTRUMENTS OF HONOR: The hoe and the shovel, the rake and harrow, are all implements of honorable and profitable industry. (Testimonies to Ministers and Gospel Workers p. 244)

TEACHERS UPHOLDING CHRIST AS THEY WORK: The teacher who has a right understanding of the work of true education, will not think it sufficient now and then to make casual reference to Christ. With his own heart warm with the love of God, he will constantly uplift the Man of Calvary. His own soul imbued with the Spirit of God, he will seek to fasten the attention of the students upon the pattern, Christ Jesus, the chiefest among ten thousand, the One altogether lovely. (Fundamentals of Christian Education p. 526)

GOOD HEALTH NEEDED BY TEACHERS: For almost every other qualification that contributes to his success, the teacher is in great degree dependent upon physical vigor. The better his health, the better will be his work. (Education p. 277)

THE MORE PERFECT THE HEALTH, THE MORE PERFECT THE LABOR: The importance of the teacher's physical qualifications can hardly be overestimated; for the more perfect his health, the more a perfect will be his labor. (Counsels to Parents, Teachers and Students p. 177)

FAITHFUL, TENDER TEACHERS NEEDED WITH A LOVE FOR YOUNG PEOPLE: Tenderness, sympathy, unity, and love are to be cherished. There should be unselfish, devoted, faithful teachers, teachers who are constrained by the love of God and who, with hearts full of tenderness, will have a care for the health and happiness of the students. (Testimonies to the Church Vol. 6 p. 152)

TEACHERS TO WAKE UP TO THE IMPORTANCE OF AGRICULTURE AND OTHER INDUSTRIES, AND WEAVE THE WORD OF GOD INTO THEIR EFFORTS: Let the teachers wake up to the importance of this subject, and teach agriculture and the other industries that it is essential for the students to understand. Let them seek in every department of labor to reach the very best results. Let the science of the word of God be brought into the work, that the students may understand correct principles and may reach the highest possible standard. (Counsels to Parents, Teachers and Students p. 315)

TEACHERS IN THE SCHOOL TO PUT THEIR BEST ENERGIES INTO THE DEVELOPMENT OF THE LORD'S FARM: Let the teachers in our school wake up, and impart the knowledge they have in agricultural lines, and in the industries that it is essential for the students to understand,—seek in every line of labor to reach the very best results. Let the science of the word of God be brought into the work, that the students may understand correct principles, and may reach the highest possible standard. Exert your God-given abilities, and bring all your energies into the development of the Lord's farm. Study and labor, that the best results and the greatest returns may come from the seed sowing, that there may be an abundant supply of food, both temporal and spiritual. (Advocate, February 1, 1899)

TALENTED TEACHERS NEEDED: The teacher must have aptness for his work. He must have the wisdom and tact required in dealing with minds. However great his scientific knowledge, however excellent his qualifications in other lines, if he does not gain the respect and confidence of his pupils, his efforts will be in vain. Teachers are needed who are quick to discern and improve every opportunity for doing good; those who with enthusiasm combine true dignity; who are able to control, and apt to teach, who can inspire thought, arouse energy, and impart courage and life. (Education pp. 278-279)

TEACHERS SHOULD BE WORKING WITH THE STUDENTS, DEVOTING A PORTION OF EACH DAY TO ACTIVE PHYSICAL LABOR: I have been led to inquire, Must all that is valuable in our youth be sacrificed in order that they may obtain an education at the schools? If there had been agricultural and manufacturing establishments in connection with our schools, and competent teachers had been employed to educate the youth in the different branches of study and labor, devoting a portion of each day to mental improvement, and a portion of the day to physical labor, there would now be a more elevated class of youth to come upon the stage of action, to have influence in moulding society. The youth who would graduate at such institutions would many of them come forth with stability of character. They would have perseverance, fortitude, and courage to surmount obstacles, and principles that would not be swerved by wrong influence, however popular. (Christian Education p. 18)

TEACHERS SHOULD WORK ALONGSIDE THE STUDENTS IN OUTDOOR LABOR: Our youth need an education in felling trees and tilling the soil as well as in literary lines. Different teachers should be appointed to oversee a number of students in their work and should work with them. Thus the teachers themselves will learn to carry responsibilities as burden bearers. Proper students also should in this way be educated to bear responsibilities and to be laborers together with the teachers. All should counsel together as to the very best methods of carrying on the work. (Testimonies to the Church Vol. 6 p. 179)

TEACHERS SHOULD WORK WITH THE STUDENTS OUT IN THE FIELDS: Let the teachers in our schools take their students with them into the gardens and fields, and teach them how to work the soil in the very best manner. (Special Testimonies on Education p. 102)

GOD WOULD BE PLEASED WHEN TEACHERS WORK A FARM INTELLIGENTLY: The farm, if worked intelligently, is capable of furnishing fruit and other produce for the school. The teachers, both in their work in the schoolroom and on the farm, should constantly seek to reach a higher standard, that they may be better able to teach the students how to care for the trees, the berries, the vegetables, and the grains that shall be raised. This will be pleasing to God and will bring the approval and respect of those in the community who understand the principles of agriculture. (A Place Called Oakwood p. 119)

ENTHUSIASM NEEDED: The teacher in his work is dealing with things real, and he should speak of them with all the force and enthusiasm which a knowledge of their reality and importance can inspire. (Education p. 233)

TEACHERS WHO SPEND TIME WITH THEIR STUDENTS OUTSIDE OF THE FORMAL CLASSROOM WILL BE MOST EFFECTIVE: In lines of recreation for the student the best results will be attained through the personal co-operation of the teacher. The true teacher can impart to his pupils few gifts so valuable as the gift of his own companionship. It is true of men and women, and how much more of youth and children, that only as we come in touch through sympathy can we understand them; and we need to understand in order most effectively to benefit. To strengthen the tie of sympathy between teacher and student there are few means that count so much as pleasant association together outside the schoolroom. In some schools the teacher is always with his pupils in their hours of recreation. He unites in their pursuits, accompanies them in their excursions, and seems to make himself one with them. Well would it be for our schools were this practice more generally followed. The sacrifice demanded of the teacher would be great, but he would reap a rich reward. (Education p. 212)

WHEN TEACHERS AND STUDENTS UNITE IN DOING AGRICULTURAL WORK, WE ARE FOLLOWING THE EXAMPLE OF THE SCHOOL OF THE PROPHETS. THIS IS GOD'S PLAN: The Lord instructed me that some connected with the institution would not see the necessity of uniting agricultural work with the instruction given in the school. In all our educational institutions physical and mental work should have been combined. In vigorous physical exercise, the animal passions find a healthy outlet and are kept in proper bounds. Healthful exercise in the open air will strengthen the muscles, encourage a proper circulation of blood, help to preserve the body from disease, and will be a great help in spirituality. For many years it has been presented to me that teachers and students should unite in this work. This was done anciently in the schools of the prophets. (Manuscript 40, 1903, p. 11)

MINISTERS WOULD DO WELL TO SPEND PART OF EACH DAY IN THE FIELDS WITH THE STUDENTS: It would be well if ministers who labor in word or doctrine could enter the fields and spend some portion of the day in physical exercise with the students. They could do as Christ did in giving lessons from nature to illustrate Bible truth. Both teachers and students would have much more healthful experience in spiritual things, and much stronger minds and purer hearts to interpret eternal mysteries, than they can have while studying books so constantly, and working the brain without taxing the muscles. God has given men and women reasoning powers, and he would have men employ their reason in regard to the use of their physical machinery. The question may be asked, How can he get wisdom that holdeth the plow, and driveth oxen?—By seeking her as silver, and searching for her as for hid treasures. For his God doth instruct him to discretion, and doth teach him. This also cometh forth from the Lord of Hosts, which is wonderful in counsel, and excellent in working. (Special Testimonies on Education p. 102)

(1) A Competent Farm Manager to be Employed

EXPERIENCED FARM MANAGEMENT NEEDED: Experience is of great value. The Lord desires to have men of intelligence connected with His work…The human agent should strive to attain perfection, that he may be an ideal Christian, complete in Christ Jesus. (Testimonies to the Church Vol. 7 pp. 247-248)

A COMPETENT FARM MANAGER TO BE EMPLOYED: A competent farm manager should be employed, also wise, energetic men to act as superintendents of the several industrial enterprises, men who will use their undivided talents in teaching the students how to work. (Testimonies to the Church Vol. 6 p. 182)

Characteristic #8: Self-sufficiency

— (1): Farming a Good Investment for Schools

DESPITE THE COST, THE OUTLAY PUT FORTH FOR MANUAL TRAINING WILL PROVE THE GREATEST ECONOMY: Even from the viewpoint of financial results, the outlay required for manual training would prove the truest economy. Multitudes of our boys would thus be kept from the street corner and the groggery; the expenditure for gardens, workshops, and baths would be more than met by the saving on hospitals and reformatories. And the youth themselves, trained to habits of industry, and skilled in lines of useful and productive labor—who can estimate their value to society and to the nation? (Education p. 218)

DESPITE THE HIGH COST, EVERY OUTLAY SPENT TOWARDS INDUSTRIAL TRAINING OF THE YOUTH IS MONEY WELL SPENT: The objection most often urged against industrial training in the schools is the large outlay involved. But the object to be gained is worthy of its cost. No other work committed to us is so important as the training of the youth, and every outlay demanded for its right accomplishment is means well spent. (Education p. 218)

PROPER CULTIVATION TOGETHER WITH GOD'S BLESSING TO YIELD SURPRISING RESULTS: No one need to have regret in reference to this land; for with proper working it will surprise the people in this section of the country. All the regret I have is that we have not money to take in sections of the land that would extend the ground. I have not one doubt in reference to the securing of this land. If the Lord prospers those who occupy it and who cultivate it, as we believe He will, we shall see a change that will surprise all who look upon it. I can hardly endure the thought that time is passing, and that the work of clearing the land is delayed. (Manuscript Release 13 p. 352)

— (2): Financial Losses during the Start-up Phase

SUCCESS DEPENDS MORE ON PERSEVERANCE THAN ON BOOK KNOWLEDGE: Serious times are before us, and there is great need for the families to get out of the cities into the country, that the truth may be carried into the highways and byways of the earth. Much depends upon your laying your plans according to the word of the Lord, and with persevering energies to go ahead. More depends upon active perseverance than upon genius and book knowledge. All the talents and ability given to human agents, if unworked, are of no value. The talent of genius must be constantly worked. Make a beginning. The tree is in the acorn, and the acorn in the tree. (Manuscript Release 8 p. 200)

THE EARTH HAS BLESSINGS FOR THOSE WITH THE COURAGE, WILL AND PERSEVERANCE TO FARM: Many are unwilling to earn their bread by the sweat of their brow, and they refuse to till the soil. But the earth has blessings hidden in her depths for those who have courage and will and perseverance to gather her treasures. Fathers and mothers who possess a piece of land and a comfortable home are kings and queens. (Special Testimonies on Education p. 104)

FINANCIAL LOSS MAY BE INCURRED DURING INITIAL PHASE OF AGRICULTURAL AND INDUSTRIAL TRAINING. THE REWARDS ARE WORTH THE EXPENSE: I urge that our schools be given encouragement in their efforts to develop plans for the training of the youth in agricultural and other lines of industrial work. When, in ordinary business, pioneer work is done and preparation is made for future development, there is frequently a financial loss. But let us remember the blessing that physical exercise brings to the students. Many students have died while endeavoring to acquire an education, because they confined themselves too closely to mental effort. (Counsels to Parents, Teachers and Students p. 317)

INITIAL STARTUP OF INDUSTRIAL TRAINING MAY RESULT IN FINANCIAL LOSS—YET THE WORK MUST BE STARTED: I urge that our other schools be given encouragement in their efforts to develop plans for the training of the youth in agricultural and other lines of industrial work. When, in ordinary business, pioneer work is done, and preparation is made for future development, there is frequently a financial loss. And as our schools introduce manual training, they, too, may at first incur loss. But let us remember the blessing that physical exercise brings to the students. Many students have died while endeavoring to acquire an education, because they confined themselves too closely to mental effort. (Pamphlet 164 p. 37)

INDUSTRIES SHOULD NOT BE DROPPED JUST BECAUSE DIFFICULTIES ARISE: Because difficulties arise, we are not to drop the industries that have been taken hold of as branches of education. While attending school the youth should have an opportunity for learning the use of tools. Under the guidance of experienced workmen, carpenters who are apt to teach, patient, and kind, the students themselves should erect buildings on the school grounds and make needed improvements, thus by practical lessons learning how to build economically. The students should also be trained to manage all the different kinds of work connected with printing, such as typesetting, presswork, and book binding, together with tentmaking and other useful lines of work. Small fruits should be planted, and vegetables and flowers cultivated, and this work the lady students may be called out of doors to do. Thus, while exercising brain, bone, and muscle, they will also be gaining a knowledge of practical life. (Testimonies to the Church Vol. 6 p. 176)

— (3): The School to Grow its Own Food

OUR SCHOOLS SHOULD GROW THEIR OWN PRODUCE: Our schools should not depend upon imported produce, for grain and vegetables, and the fruits so essential to health. (Testimonies to the Church Vol. 6 p. 179)

SCHOOLS SHOULD GROW THEIR OWN PRODUCE: This institution must not depend upon imported produce,—for the fruits so essential to healthfulness, and for their grains and vegetables. This is the very first work that must be entered upon. Then as we shall advance and add to our facilities, advance studies and object lessons should come in. We are not to subtract from that which has already been taken hold of as a branch of education. (Spalding and Magan p. 134)

Chapter #5: Characteristics of Model Schools that Follow God's Plan

THE SCHOOL HAD ITS OWN LAND AND ORCHARD FOR GROWING FOOD: Our school is located here. Their land was cleared and planted with trees at the same time that my orchard was planted. This coming season we expect that it will bear fruit for the school. Our people are settling in this place. Here students are to be educated in books, and are also to be taught how to do all kinds of manual labor. The Lord will help us in this work. This is the first term of school. There are sixty students in attendance. Thirty of these come from a distance, and live in the home. All the students are young men and young women of excellent capabilities. (Manuscript Release 7 p. 253)

A SCHOOL THAT PLANNED TO RAISE ITS OWN FRUITS AND VEGETABLES: Today Mr. Moseley comes to bring oranges and lemon trees for us to set out. As soon as this work is done, we shall begin to plant vegetables. We have to get our groceries from Sydney, nearly a hundred miles away, or from Newcastle, twenty-two miles. But we hope soon to raise our own fruit and vegetables. (Letter 42, 1895, to J. H. Kellogg)

EXAMPLE OF A MODEL SCHOOL: Spread out before me (at Avondale) I saw land planted with every kind of fruit tree that will bear fruit in this locality; there were also vegetable gardens, where seeds were sown and cultivated. (Testimonies to the Church Vol. 6 p. 185)

AN ABUNDANT SUPPLY OF FOOD: Study and labor, that the best results and the greatest returns may come from the seed sowing, that there may be an abundant supply of food, both temporal and spiritual, for the increased number of students that shall be gathered in to be trained as Christian workers. (Testimonies to the Church Vol. 6 p. 192)

FRESH FRUITS FOR HEALTH: The Lord desires those living in countries where fresh fruit can be obtained during a large part of the year, to awake to the blessing they have in this fruit. The more we depend upon the fresh fruit just as it plucked from the tree, the greater will be the blessing. (Testimonies to the Church Vol. 7 p. 186)

FRESH FRUIT TO BE EATEN OF FREELY: It would be well for us to do less cooking and to eat more fruit in its natural state. Let us teach the people to eat freely of the fresh grapes, apples, peaches, pears, berries, and all kinds of fruit that can be obtained. (Testimonies to the Church Vol. 7 p. 134)

FRUITS USED IN WINTER SAUCES: Pears and cherries, if they can be obtained, make very nice sauce for winter use. (Counsels on Diet and Foods p. 312)

APPLES A SUPERIOR FRUIT: If you can get apples, you are in good condition as far as fruit is concerned, if you have nothing else. Apples are superior to any fruit for a standby that grows. (Counsels on Diet and Foods p. 312)

— (4): Farming a Means of Financial Support for the School

VARIOUS INDUSTRIES SHOULD BE STARTED TO MAKE THE SCHOOL SELF-SUPPORTING: The schools in Graysville and Huntsville were established in the order of God. They are to do a work for Him. They are to become self-supporting, by making the best use of their land, by raising those products best suited to the climate and soil of their locality. Various industries are to be established. The Lord will greatly bless these industries if the workers will walk in His counsel. If they will look to Him, He will be their wisdom and their righteousness. His wisdom will be seen in the work of those who follow His directions. He will teach all who will learn of Him His meekness and lowliness. (General Conference Bulletin, April 14, 1903)

FARMING TO BE A MEANS OF SUPPORTING THE SCHOOLS: The plan upon which our brethren propose to work is to select some of the best and most substantial young men and women from Berrien Springs and other places in the North, who believe that God has called them to the work in the South, and give them a brief training as teachers. Thorough instruction will be given in Bible study, physiology, and the history of our message; and special instruction in agriculture will be given. It is hoped that many of these students will eventually connect with schools in various places in the South. In connection with these schools there will be land that will be cultivated by teachers and students, and the proceeds from this work will be used for the support of the schools. (Review and Herald, August 18, 1904)

VARIOUS INDUSTRIES SHOULD BE STARTED TO MAKE THE SCHOOL SELF-SUPPORTING: The schools in Graysville and Huntsville were established in the order of God. They are to do a work for Him. They are to become self-supporting, by making the best use of their land, by raising those products best suited to the climate and soil of their locality. Various industries are to be established. The Lord will greatly bless these industries if the workers will walk in His counsel. If they will look to Him, He will be their wisdom and their righteousness.

Chapter #5: Characteristics of Model Schools that Follow God's Plan

His wisdom will be seen in the work of those who follow His directions. He will teach all who will learn of Him His meekness and lowliness. (General Conference Bulletin, April 14, 1903)

AGRICULTURE TO BE A FINANCIAL BENEFIT TO OUR SCHOOLS: In establishing schools, one important point is to secure land sufficient for the carrying forward of industries that will enable the students to be self-supporting. There should be land sufficient for the raising of the fruit and vegetables required by the school, and also some for sale. Agriculture should be made a financial benefit to the school. (Pamphlet 151 p. 76)

EXAMPLE OF THE SCHOOLS OF THE PROPHETS: To meet this growing evil, God provided other agencies to aid parents in the work of education. From the earliest times, prophets had been recognized as teachers divinely appointed. In the highest sense the prophet was one who spoke by direct inspiration, communicating to the people the messages received from God. But the name prophet was given also to those who, though not so directly inspired, were divinely called to instruct the people in the works and ways of God. For the training of such a class of teachers, Samuel, by the Lord's direction, established the schools of the prophets. These schools were intended to serve as a barrier against the widespread corruption, to provide for the mental and spiritual welfare of the youth, and to promote the prosperity of the nation by furnishing it with men qualified to act in the fear of God as leaders and counselors. To this end, Samuel gathered companies of young men who were pious, intelligent, and studious. These were called the sons of the prophets. As they studied the Word and works of God, His life-giving power quickened the energies of mind and soul, and the students received wisdom from above. The instructors were not only versed in divine truth, they had themselves enjoyed communion with God, and had received the special endowment of His Spirit. They had the respect and confidence of the people, both for learning and for piety. In Samuel's day there were two of these schools—one at Ramah, the home of the prophet, the other at Kirjathjearim. Later others were established. The pupils of these schools sustained themselves by their own labor in tilling the soil or in some mechanical employment. Many, also, of the teachers supported themselves by manual labor. In Israel it was regarded as a sin to allow children to grow up in ignorance of useful labor. (True Education p. 32)

BOUNTY AT AVONDALE: Has not the Lord blessed? From one of the reports received, we learn that last year seven thousand pounds of honey of the best quality has been made on the school estate. Large quantities of vegetables have been raised, and the sale of the surplus has been a source of considerable revenue to the school. All this is very encouraging to us; for we took the wild land, and helped to bring it to its present fruitful state. To the Lord we ascribe all the praise. (Selected Messages Vol. 1 p. 102-103)

PROCEEDS FROM FOOD GROWN TO HELP SUPPORT THE SCHOOL: In connection with these schools there will be land that will be cultivated by teachers and students, and the proceeds from this work will be used for the support of the schools. (Special Testimonies, Series B, number 11)

AN EXPERIENCE RECORDED FOR OUR ENCOURAGEMENT: In the darkest hour of the establishment of the Avondale School, when the outlook seemed the most discouraging, I was sitting in the hotel in Cooranburg then used by our people, completely wearied out by the complaints made regarding the land. My heart was sick and sore. But suddenly a great peace came upon me. Angels were in the room, and then the words were spoken, 'Look ye,' and I saw flourishing cultivated land, bearing its cream of fruit and root crops. Many resources were spread out before me, and wherever my eye was directed, I saw prosperity. I saw the school filled with promising students. All seemed to be helped, by the inspiration of well organized efforts, to stand and work upon a high platform. There was so large a number of pleasant faces that I could not fail to understand that the light of the Lord's countenance was lifted upon them. A great light and peace came upon me. I was so blessed that I praised the Lord aloud, saying, 'His word is fulfilled, God will spread a table in the wilderness. Since then I have not had one hour of discouragement in regard to the Avondale school. (Letter 36, 1907, written from Sanitarium, California, to Elder Irwin)

KNOW THAT AN AGRICULTURAL PROGRAM CAN STRENGTHEN A SCHOOL FINANCIALLY: Schools are to be established away from the cities, where the youth can learn to cultivate the soil and thus help to make themselves and the school self-supporting. In connection with these schools all the different lines of work, whether agricultural or mechanical, that the situation of the place will warrant are to be developed. Let means be gathered for the establishment of such schools. In them students may gain an education that, with God's blessing, will prepare them to win souls to Christ. If they unite with the Savior they will grow in spirituality and will become valuable workers in His vineyard. (Testimonies to the Church Vol. 7 p. 232)

Chapter #5: Characteristics of Model Schools that Follow God's Plan
— (5): Why Some Farms Do Poorly

CHEAP, SUPERFICIAL FARMING WILL RESULT IN COMMENSURATE HARVESTS: We walked over one farm where the land had been cleared, and which joined the school land. We examined the way in which they work the land, and found that the plough had been put in only to about the depth of six inches. An intelligent American farmer would not regard this as a faithful way of working the land. Those who work in this cheap, superficial way cannot expect to receive anything out of harmony with their method, but in accordance with it. (Manuscript Release 13 p. 350)

MANY FARMS DO POORLY DUE TO WRONG ATTITUDES, NEGLECT AND A FOCUS ON ENTERTAINMENT: Many farmers have failed to secure adequate returns from their land because they have undertaken the work as though it was a degrading employment; they do not see that there is a blessing in it for themselves and their families. All they can discern is the brand of servitude. Their orchards are neglected, the crops are not put in at the right season, and a mere surface work is done in cultivating the soil. Many neglect their farms in order to keep holidays and to attend horse-races and betting clubs; their money is expended in shows and lotteries and idleness, and then they plead that they cannot obtain money to cultivate the soil and improve their farms; but had they more money, the result would still be the same. (Special Testimonies on Education p. 105)

GROUND MUST BE PREPARED: Remember that you can have no harvest unless the ground is properly prepared for the seed; failure may be wholly due to neglect on this point. (Fundamentals of Christian Education p. 323)

SURFACE WORK AND NEGLECT THE REASON FOR FAILURE ON MANY FARMS: Many who till the soil fail to secure adequate returns because of their neglect. Their orchards are not properly cared for, the crops are not put in at the right time, and a mere surface work is done in cultivating the soil. Their ill success they charge to the unproductiveness of the land. False witness is often borne in condemning land that, if properly worked, would yield rich returns. The narrow plans, the little strength put forth, the little study as to the best methods, call loudly for reform. (Ministry of Healing p. 193)

IF MONEY AND TIME SPENT ON AMUSEMENT WERE PUT TOWARDS EDUCATION AND INTELLIGENT FARMING, WE WOULD SEE A DIFFERENT

Hope in the Soil

STATE OF THINGS: Now the case where a man owns his place clear is a happy exception to the rule. Merchants are failing, families are suffering for food and clothing. No work presents itself. But the holidays are just as numerous. Their amusements are entered into as eagerly. All who can do so will spend their hard-earned pence and shillings and pounds for a taste of pleasure, for strong drink, or some other indulgence. The papers that report the poverty of the people, have regular standing notices of the horse-races, and of the prizes presented for different kinds of exciting sports. The shows, the theaters, and all such demoralizing amusements, are taking the money from the country, and poverty is continually increasing. Poor men will invest their last shilling in a lottery, hoping to secure a prize, and then they have to beg for food to sustain life, or go hungry. Many die of hunger, and many put an end to their existence. The end is not yet. Men take you to their orchards of oranges and lemons, and other fruits, and tell you that the produce does not pay for the work done in them. It is next to impossible to make ends meet, and parents decide that the children shall not be farmers; they have not the courage and hope to educate them to till the soil. What is needed is schools to educate and train the youth so that they will know how to overcome this condition of things. There must be education in the sciences, and education in plans and methods of working the soil. There is hope in the soil, but brain and heart and strength must be brought into the work of tilling it. The money devoted to horse-racing, theater-going, gambling, and lotteries; the money spent in the public houses for beer and strong drink,— let it be expended in making the land productive, and we shall see a different state of things. (Special Testimonies on Education p. 94)

IDLENESS AND FOCUS ON AMUSEMENTS CONTRIBUTES TO FAILURE IN FARMING: We should judge that the general difficulty with farming here is a lack of interest. There is plenty of idleness, [with] numerous holidays which are improved in following many kinds of objectionable amusements. The people are interested in horse-racing and card playing, in smoking and drinking, and this kind of employment benefits neither themselves nor others. They pass away their time in this way, and the lands are neglected. But if the soil were cultivated, it would produce excellent fruit. (Manuscript Release 13 p. 349)

SUPERFICIAL PLOWING RESULTS IN MEAGER HARVEST: I do not mean that a superficial work should be done as is illustrated by the way in which some portions of the land are worked in Australia. The plow was only put in the depth of a few inches, the ground was not prepared for the seed, and the harvest was meager, corresponding to the superficial preparation that was given to the land. (Special Testimonies on Education p. 222)

Chapter #5: Characteristics of Model Schools that Follow God's Plan

Characteristic #9: Obedience to God's Plan Vital to Farming Success

IF FARM MANAGERS AND TEACHERS ALLOW THE HOLY SPIRIT TO WORK WITH THEM, GOD WILL BLESS THEIR WORK AND THE FARM: If the managers of this farm and the teachers in the school will receive the Holy Spirit to work with them, they will have wisdom in their management, and God will bless their labors. The care of the trees, the planting and the sowing, and the gathering of the harvest are to be wonderful lessons for all the students. The invisible links which connect the sowing and the reaping are to be studied, and the goodness of God is to be pointed out and appreciated. It is the Lord that gives the virtue and the power to the soil and to the seed. Were it not for the divine agency, combined with human tact and ability, the seed sown would be useless. There is an unseen power constantly at work in man's behalf to feed and to clothe him. The parable of the seed as studied in the daily experience of teacher and student is to reveal that God is at work in nature, and it is to make plain the things of the kingdom of heaven. (Testimonies to the Church Vol. 6 p. 185)

GOD WILL INSTRUCT THOSE WHO PRAY AND FOLLOW HIS COUNSEL ON WHAT THEY SHOULD DO: To all who would mark out a certain definite course for their brother to pursue, the Lord says, Stand out of the way. Satan and his emissaries are doing enough of this kind of work. We are altogether too near the close of this earth's history to seek to block the wheels of the chariot of truth. God's workers are to come into line, to pray together, to counsel together. And whenever it is impossible for them to gather for counsel, God will instruct through His Spirit those who sincerely desire to serve Him. (Manuscript Release 5 p. 280)

MEN MOURN OVER POOR CROP YIELDS AND CONDEMN LAND WHICH WOULD PRODUCE RICH YIELDS IF BIBLICAL PRINCIPLES WERE FOLLOWED: False witness has been borne in condemning land which, if properly worked, would yield rich returns. The narrow plans, the little strength put forth, the little study as to the best methods, call loudly for reform. The people need to learn that patient labor will do wonders. There is much mourning over unproductive soil, when if men would read the Old Testament Scriptures they would see that the Lord knew much better than they in regard to the proper treatment of land. After being cultivated for several years, and giving her treasure to the possession of man, portions of the land should be allowed to rest, and then the crops should be changed. (Special Testimonies on Education p. 100)

DIVINE HELP NEEDED TO BRING TREASURES FROM THE EARTH: God has originated and proclaimed the principles on which divine and human agencies are to combine in temporal matters as well as all spiritual achievements. They are to be linked together in all human pursuits, in mechanical and agricultural labor, in mercantile and scientific enterprises. In all lines of work it is necessary that there be co-operation between God and man. God has provided facilities with which to enrich and beautify the earth. But the strength and ingenuity of human agencies are required to make the very best use of the material. God had filled the earth with treasure, but the gold and silver are hidden in the earth, and the exercise of man's powers is required to secure this treasure which God has provided. Man's energy and tact are to be used in connection with the power of God in bringing the gold and silver from the mines, and trees from the forest. But unless by his miracle-working power God co-operated with man, enabling him to use his physical and mental capabilities, the treasures in our world would be useless. (Review and Herald, May 28, 1908)

GOD CAN MAKE SEEDS THAT WERE THROWN AWAY INTO LIVING PLANTS. MAN MUST COOPERATE WITH GOD TO GARNER THE HARVEST: He [God] employs many unseen agencies to make the seeds apparently thrown away, living plants. First appear the blade, then the ear, then the full corn in the ear. God creates the electricity that gives life to the seed, vitality to the blade, the ear, and the corn in the ear. Who else can be depended on to give the due proportion required of all the agencies to perfect the harvest of fruits and grains? Let man employ his agencies to the utmost limit; he must then depend on his Creator, who knows just what is needed for the harvest, which is connected to Him by wonderful links of His own wonderful power, beyond the human agency. Without these unseen agencies, seed is valueless. (Manuscript Release 3 p. 322)

GOD IS THE MANAGER, THE HOUSEHOLDER AND CONTROLLER OF NATURE. ALL HARVESTS RESULT FROM HIS BLESSINGS: This was to show that nature was not God, that God controlled nature. God designed that from nature His church should constantly learn important lessons. They were to cherish a vivid sense that God was the manager, the householder. They were to know the reality of His presence and His providential care over all the earth. They were to realize that all nature was under His supervision, all the productions of the ground under His ministration. This was to give them faith in His providence. He could withhold His blessings or bestow them. (Manuscript Release 3 p. 347)

Chapter #5: Characteristics of Model Schools that Follow God's Plan

GOD TO BE OUR INSTRUCTOR ON HOW TO EFFECTIVELY GARDEN: He who taught Adam and Eve in Eden how to tend the garden, would instruct men today. There is wisdom for him who holds the plow, and plants and sows the seed. (<u>Special Testimonies on Education</u> p. 103)

GOD WILL BLESS THE LABORS OF DILIGENT, GODLY FARMERS, ENABLING THEM TO EARN A LIVING FROM THE SOIL: This country needs educated farmers. The Lord gives the showers of rain and the blessed sunshine. He gives to men all their powers; let them devote heart and mind and strength to doing his will in obedience to his commandments. Let them cut off every pernicious habit, never expending a penny for beer or liquor of any kind, nor for tobacco, having nothing to do with horse-racing or similar sports, and then commit themselves to God, working with their endowment of physical strength, and their labor will not be in vain. That God who has made the world for the benefit of man, will provide means from the earth to sustain the diligent worker. The seed placed in thoroughly prepared soil, will produce its harvest. God can spread a table for his people in the wilderness. (<u>Special Testimonies on Education</u> p. 95)

AS LONG AS THE CHILDREN OF ISREAL FOLLOWED GOD'S PLAN FOR THE LAND THERE WERE NO BEGGARS AMONG THEM: People were subject to misfortune, sickness, and loss of property the same then [ancient Israel] as now; but so long as they followed the instruction given by God there were no beggars among them, neither any who suffered for food. Their wise Governor, foreseeing that misfortune would befall some, made provision for them. When the people entered Canaan, the land was divided among them according to their numbers, and special laws were enacted to prevent any one person from joining field to field, and claiming as his, all the land that he desired, or had money to purchase. No one was allowed to choose the most fertile parts for himself, and leave the poor and less desirable portions for his brother; for this would cultivate selfishness and a spirit of oppression, and give cause for dissatisfaction, complaint, and dissension. (<u>Historical Sketches</u> p. 164)

HAD OUR SCHOOLS PLACED MORE EMPHASIS ON AGRICULTURAL LINES, THEY WOULD BE MUCH MORE SUCCESSFUL: Had all our schools encouraged work in agricultural lines, they would now have an altogether different showing. There would not be so great discouragements. Opposing influences would have been overcome; financial conditions would have changed. With the students, labor would have been equalized; and as all the human machinery was proportionately taxed, greater physical and mental strength would have been developed. But the instruction which the Lord has been pleased to give has been taken hold of so feebly that obstacles have not been overcome. (Testimonies to the Church Vol. 6 p. 177)

WE ARE UTTERLY DEPENDENT ON GOD: We cannot keep ourselves for one moment. We are kept by the power of God through faith unto salvation. We are utterly dependent upon God every moment of our lives. (Review and Herald, May 28, 1908)

WITHOUT GOD'S BLESSING, NO SEEDS WOULD GROW. WITHOUT MAN'S PLANTING, NO GRAIN WOULD BE PLANTED. MAN MUST PUT FORTH HIS BEST EFFORTS, THEN DEPEND ON GOD: The parable of the seed reveals that God is at work in nature. The seed has in itself a germinating principle, a principle that God Himself has implanted; yet if left to itself the seed would have no power to spring up. Man has his part to act in promoting the growth of the grain. He must prepare and enrich the soil and cast in the seed. He must till the fields. But there is a point beyond which he can accomplish nothing. No strength or wisdom of man can bring forth from the seed the living plant. Let man put forth his efforts to the utmost limit, he must still depend upon One who has connected the sowing and the reaping by wonderful links of His own omnipotent power. (Christ's Object Lessons p. 63)

WITHOUT GOD'S HELP AND COOPERATION, GROWING WILL NOT BE SUCCESSFUL: God gives to us, that we may give. He desires us to be laborers together with him. In heaven he is carrying forward the great work of redemption. That work engages the divine councils. It requires the ministry of angels upon the earth; and it requires also our co-operation. In the natural world, man must do his part in the work of the earth. He must till and prepare the soil. And God, working through nature, giving sunshine and showers, quickens the seed sown, and causes vegetation to flourish. Thus the sowing is rewarded in the reaping of earth's treasures in bountiful harvests. The lesson is true in spiritual as in temporal things. Man must work under the guidance of the divine hand; for unless God co-operates with him, there will be no increase. Human power cannot cause the seed sown to spring into life. But there can be no reaping unless the human hand acts its part in the sowing of the seed. (Review and Herald, December 8, 1896)

COOPERATION WITH GOD He that goeth forth and weepeth, bearing precious seed, shall doubtless come again with rejoicing, bringing his sheaves with him. Psalm 126:6. Our heavenly Father gives the rain, the dew, and the sunshine from heaven to refresh the flowers and to cause vegetation to spring up and flourish. But man has a part to act, to prepare the soil and to put the seeds into the ground in order to have a harvest. If he had folded his arms and said, I will let

Chapter #5: Characteristics of Model Schools that Follow God's Plan

things take their course.... God will give the harvest. He will give the sunshine and the rain from heaven, and I will take my ease, what kind of harvest would come? Man must cooperate with God and act his part in preparing the soil and in sowing the seed, and God will give the increase. Our heavenly Father has not sent angels from heaven to preach salvation to men. He has opened to us the precious truths of His Word and implanted the truth in our hearts that we may give it to those who are in darkness. If we have indeed tasted of the precious gifts of God in His promises, we are to impart this knowledge to others.... We are individually to work as though a great responsibility rested upon us. We are to manifest untiring energy and tact and zeal in this work and take the burden, feeling the peril in which our neighbors and friends are placed. We are to work as Christ worked. We are to present the truth as it is in Jesus, that the blood of souls shall not be upon our garments. At the same time we are to feel entire dependence and trust in God, for we know we cannot do anything without His grace and power to help. A Paul may plant, and an Apollos water, but God alone can give the increase. Then we are indeed to go forward to the work, weeping, sowing the precious seeds of truth and trusting in God to give the increase. (In Heavenly Places p. 331)

WITHOUT GOD'S HELP, EVERY HUMAN EFFORT IS WORTHLESS: God desires every human being in our world to be a worker together with him. This is the lesson we are to learn from all useful employment, making homes in the forest, felling trees to build houses, clearing land for cultivation. God has provided the wood and the land, and to man he has given the work of putting them in such shape that they will be a blessing. In this work man is wholly dependent upon God. The fitting of the ships that cross the broad ocean is not alone due to the talent and ingenuity of the human agent. God is the great Architect. Without his co-operation, without the aid of the higher intelligences, how worthless would be the plans of men. God must aid, else every device is worthless. (Review and Herald, May 28, 1908)

BY COOPERATING WITH HIS EARTHLY FATHER IN FARMING, ELISHA LEARNED TO COOPERATE WITH GOD: The son of a wealthy farmer, Elisha had taken up the work that lay nearest. While possessing the capabilities of a leader among men, he received a training in life's common duties. In order to direct wisely, he must learn to obey. By faithfulness in little things, he was prepared for weightier trusts. Of a meek and gentle spirit, Elisha possessed also energy and steadfastness. He cherished the love and fear of God, and in the humble round of daily toil he gained strength of purpose and nobleness of character, growing in divine grace and knowledge. While cooperating with his father in the home duties, he was learning to cooperate with God. (Reflecting Christ p. 336)

BOTH PLANT AND SPIRITUAL GROWTH ATTAINED THROUGH COOPERATION WITH DIVINE AGENCIES: The plant grows by receiving that which God has provided to sustain its life. So spiritual growth is attained through cooperation with divine agencies. As the plant takes root in the soil, so we are to take root in Christ. As the plant receives the sunshine, the dew, and the rain, so we are to receive the Holy Spirit. If our hearts are stayed upon Christ, He will come unto us as the rain, as the latter and former rain unto the earth. As the Sun of righteousness, He will arise upon us with healing in His wings. We shall grow as the lily. We shall revive as the corn, and grow as the vine. (Hosea 6:3; Malachi 4:2; Hosea 14:5, 7) (God's Amazing Grace p. 197)

Characteristic #10: Small Schools—Connected with Small Sanitariums

CONNECT SANITARIUMS WITH SCHOOLS: In every place where schools are established we are to study what industries can be started that will give the students employment. Small sanitariums should be established in connection with our larger schools, that the students may have opportunity to gain a knowledge of medical missionary work. This line of work is to be brought into our schools as part of the regular instruction. (Medical Ministry p. 323)

SANITARIUMS AS SMALL TRAINING SCHOOLS: The sanitariums established in the future are not to be immense, expensive buildings. Small local sanitariums are to be established in connection with our schools. Many sanitariums are to be established in places outside the cities. Connected with them there are to be men and women of ability and consecration, who will conduct themselves in the love and fear of God. These institutions are to be training schools. (Medical Ministry p. 156)

KEEP SCHOOLS AND SANITARIUMS SMALL: The Lord is certainly opening the way for us as a people to divide and subdivide the companies that have been growing too large to work together to the greatest advantage. And this dividing should be done, not only that the students may have greater advantages, but that the teachers may be benefited, and life and health spared. To establish another school will be better than further enlargement of the school at ___. Let another locality have the advantage of one of our educational institutions. Secure for it the best talent, and guard against the dangers of an overcrowded school. (Healthy Ministry p. 32)

LARGE INSTITUTIONS NOT NEEDED: Never, never build mammoth institutions. Let these institutions be small, and let there be more of them, that the work of winning souls to Christ may be accomplished. (Medical Ministry p. 323)

SMALL SCHOOLS MORE OF A BLESSING: Smaller schools, conducted after the plan of the schools of the prophets, would be a far greater blessing. The money which was invested in enlarging Battle Creek College to accommodate the minister's school would better have been invested in establishing schools in rural districts in America and in the regions beyond. (Testimonies to the Church Vol. 6 p. 137)

Chapter #6: a School Farms that Followed God's Plan and Prospered

God's Farm at Madison College

THE SCHOOL IN MADISON WAS ESTABLISHED AFTER GOD'S PLAN, WITH THE OPPORTUNITY TO ACHIEVE FAR-REACHING RESULTS: The Lord had directed Brethren Sutherland and Magan, men of sound principles, to establish the work at Madison. They have devised and planned and sacrificed in order to carry the work there after God's order; but the work has been long in coming to completion. It is the privilege of these brethren to receive gifts from any of our people whom the Spirit of the Lord impresses to help. They should have means—God's means—with which to do the Lord's work. . . . The Lord selected the farm at Madison, and He signified that it should be worked on right lines, that others, learning from the workers in Madison might take up a similar work and conduct it in a like manner. Brethren Sutherland and Magan are chosen of God and faithful, and the Lord of heaven says of them, I have a work for these men and women for missionary fields. The Spirit of the Lord is with His workers. He has not restricted the labors of these self-denying, self-sacrificing men. (Manuscript Release 5 p. 279)

THE SCHOOL IN MADISON WAS ESTABLISHED ALONG THE SAME LINES AS THE SCHOOL IN AVONDALE, AND GOD BLESSED: The Lord has given to the Southern field object lessons of different kinds. The education being given to the students at Madison, which trains the youth to build, to cultivate the land, and to care for cattle and poultry, will be of great advantage to them in the future. There is no better way of keeping the body in health than to follow the plan of training that the Madison school is carrying out. This is the same kind of work as we were instructed to do when we purchased the land for our school in Australia. The students had their hours for study and their hours for work on the land. They were taught to fell trees, to plant orchards, to cultivate the soil, and to erect buildings, and this training was a blessing to all who engaged in it. (Manuscript Release 11 p. 181)

DURING THE DIFFICULT DAYS AT THE TIME OF THE END THOSE WHO HAVE RECEIVED AN ALL-ROUND TRAINING SUCH AS CAN BE HAD AT MADISON WILL HAVE THE ADVANTAGE: There is plenty of land lying waste in the South that might have been improved as the land about the Madison School has been improved. The time is soon coming when God's people, because of persecution, will be scattered in many countries. Those who have received an all-round education will have the advantage where they are. The Lord

Chapter #6: a School Farms that Followed God's Plan and Prospered

reveals divine wisdom in thus leading His people to the training of all their faculties and capabilities for the work of disseminating truth… (Manuscript Release 5 p. 280)

IF MORE SCHOOLS TRAINED PRACTICAL MISSIONARIES LIKE MADISON COLLEGE DID, OUR WORK WOULD BE A SPECTACLE TO THE WORLD—AND THE GOSPEL WOULD QUICKLY BE CARRIED TO OTHER COUNTRIES: If many more in other schools were receiving a similar training, we as a people would become a spectacle to the world, to angels, and to men. The message would quickly be carried to every country, and souls now in darkness would be brought to the light. These men under the special light the Lord has given, are not to be hindered in any way, for the Lord is leading them. (Manuscript Release 5 p. 280)

THOUGH ESTABLISHED WITH GREAT DIFFICULTIES, MADISON COLLEGE PROSPERED AND WAS BLESSED BY GOD: The Lord in His providence has brought about the establishment of the Madison school through the efforts of Brethren [E. A.] Sutherland and [P. T.] Magan, and a few faithful associates. Their labors have been performed under no ordinary circumstances. These men had an experience at Berrien Springs which was a severe one, but the Lord brought them safely through it and made it a means of blessing to them. They felt that they must go to the South and labor for this needy field. They went out not knowing whither they were going, and the Lord guided them to Madison, a beautiful place of 400 acres. For a time the way for the establishment of the work seemed hedged up. The Lord led His servants through a trying experience, but He saw the end from the beginning. When some of their brethren expostulated and labored to discourage them, the Lord encouraged. And the results of the efforts put forth at that place we can see; The Lord's blessing has rested upon their efforts. (Manuscript Release 11 p. 182)

MADISON COLLEGE OFFERED THE BEST ALL-ROUND EDUCATION IN AMERICA: The work that the laborers have accomplished at Madison has done more to give a correct knowledge of what an all-round education means than any other school that has been established by Seventh-day Adventists in America. The Lord has given these teachers in the South an education that is of highest value, and it is a training that God would be pleased to have all our youth receive. (Manuscript Release 11 p. 182)

MORE SCHOOLS SHOULD BE ESTABLISHED LIKE MADISON COLLEGE: It would have been pleasing to God, if, while the Madison school has been doing its work, similar schools had been established in different parts of the Southern field. . . . (Manuscript Release 5 p. 280)

God's Farm at Avondale College

WELL-ROUNDED PHYSICAL AND SPIRITUAL EDUCATION THE GOALS OF AVONDALE COLLEGE: The school [Avondale College] was established at a great expense, both of time and labor, to enable students to obtain an all-round education, that they might gain a knowledge of agriculture, a knowledge of the common branches of education, and above all, a knowledge of the Word of God. . . . (Manuscript Release 11 p. 173)

AVONDALE WAS A MODEL SCHOOL, PATTERNED AFTER GOD'S PLAN: God will bless those schools that are conducted according to His design. When we were laboring to establish the educational work in Australia, the Lord revealed to us that this school must not pattern after any schools that had been established in the past. This was to be a sample school. It was organized on the plan that God had given us, and He has prospered its work. (Counsels to Parents, Teachers and Students p. 533)

DETAILS OF THE AVONDALE SCHOOL: Two plain, simple, substantial buildings have been erected for school purposes. The main building is not yet built. We are using a wing, which will answer until we can get means to advance on the main building. We will soon be compelled to build a chapel. We are so thankful that we have been able to make a beginning: and we earnestly desire to have this school such as the Lord shall approve. The school commences at nine o'clock in the morning, and closes at one. Then comes the dinner hour, and then three hours of physical labor; for the mental and physical powers must be proportionately taxed. (Manuscript Release 7 p. 254)

THE SCHOOL ORCHARD IN AUSTRALIA DID EXCELLENTLY WELL: The school orchard is doing excellently well. If the land is worked it will yield its treasures, but weeds will grow and those who own land will not exercise ambition to take these weeds out by the roots and give them no quarter. Deep plowing must be done. They let a few orange trees grow in the sod, also the lemons. We get the choicest, best oranges for three pence and two pence ha'penny [half penny] per dozen—six cents American money, and four and five cents per dozen for large, beautiful, sweet oranges. (Manuscript Release 8 p. 253)

GOD'S BLESSING ON AVONDALE SCHOOL: We saw the great need for a school in which promising young men and young women could be trained for the Master's service; and we went right into the woods in New South Wales, purchased fifteen hundred acres of land, and there established a training school away from the cities. Three years ago we returned to America. Others were sent to Australia to take our places. The work has continued to grow; prosperity has attended every effort. I wish you could read the letters that come to us. Doubtless you have heard of the dreadful drought that has caused famine in so many places in Australia during the past two years. Hundreds of thousands of sheep and cattle and horses have perished. In all the colonies, and especially in Queensland, the suffering and financial loss have been great. But the spot that was chosen for our training school, has had sufficient rainfall for good pasture land and bountiful crops; in fact, in legislative assemblies and in the newspapers of the great cities it has been specified as 'the only green spot in all New South Wales.' Is not this remarkable? Has not the Lord blessed? From one of the reports received, we learn that last year seven thousand pounds of honey of the best quality has been made on the school estate. Large quantities of vegetables have been raised, and the sale of the surplus has been a source of considerable revenue to the school. All this is very encouraging to us; for we took the wild land, and helped to bring it to its present fruitful state. To the Lord we ascribe all the praise. (Selected Messages Vol. 1 pp. 102-103)

STERN WORK AT AVONDALE—FOLLOWED BY THE BLESSING OF GOD: We had stern work to do in Australia in educating parents and youth along these lines; but we persevered in our efforts until the lesson was learned that in order to have an education that was complete, the time of study must be divided between the gaining of book knowledge and the securing of a knowledge of practical work. Part of each day was spent in useful work, the students learning how to clear the land, how to cultivate the soil and to build houses in the time that would otherwise have been spent in games and in seeking amusement. And the Lord blessed the students who thus devoted their time to learning lessons of usefulness. (Adventist Home pp. 508-509)

SHOULD THE LORD PROSPER THE FARM, ELLEN WHITE EXPECTED A GREAT YIELD IN AUSTRALIA: If the Lord prospers us next year, as He has done the past year, we will have all the fruit we wish to take care of, early and late. The early fruit comes when there is nothing else, so this is an important item. The peaches are rich and juicy and grateful to the taste. We have quince trees set out, and lemon, orange, apple, plum, and persimmon trees. We have even planted elderberry bushes. We planted our vineyard in June. Everything is flourishing and we shall have many clusters of grapes this season. (Manuscript Release 8 p. 252)

CROPS IN AUSTRALIA VERY SUCCESSFUL: Our crops were very successful. The peaches were the most beautiful in coloring, and the most delicious in flavor of any that I had tasted. We grew the large yellow Crawford and other varieties, grapes, apricots, nectarines, and plums. (Manuscript Release 15 p. 54)

NOT WITHOUT DIFFICULTY: The struggle it has taken to carry out what God has plainly revealed should be done, has been terrible. Satan has contested every inch of the ground; but God has given us many victories. He has planted the Avondale school, and we have the plainest evidences that He will be glorified by it. He has given minute instruction regarding its location, object, and management. Now He is telling us that if we will walk in the light He has given, Avondale will become the training ground for many missionary fields. The hand of God is in these things. (Divine Predictions, pp. 346-347)

REJOICING OVER AVONDALE SCHOOL: If I had to move from here (Elmshaven), I should want to go to Cooranbong. As I read of the fearful drought in Australia, and of the loss of life and property resulting from it, and then hear of the prosperity attending the Avondale school farm, I praise the Lord. How fresh in my mind are the words spoken by One of Authority, as some were presenting the objectionable features of the school land: 'Can not God spread a table in the wilderness?' He certainly has done this by blessing the orchard and the vegetable garden. The abundance with which the school land has produced its treasure testifies that God's prospering hand is with His people. I am as thankful for this as though I were still there. I thank the Lord in behalf of my brethren and sisters in Australia. Not one thing has failed of that which He has promised. Let us take courage, and rejoice in the Lord. (Letter B-9, 1903, written from Elmshaven in January of that year)

Chapter #6: a School Farms that Followed God's Plan and Prospered

— (1) A Letter from Avondale

Cooranbong, Australia,
August 27, 1895
Dear Brother and Sister -----,

The students of our manual training school at this place are doing their best to follow the light God has given to combine with mental training the proper use of brain and muscle. Thus far the results have exceeded our expectations. At the close of the first term, which was regarded as an experiment, opportunity was given for the students to have their vacation, and engage in whatever work they chose to do. But everyone begged that the school might be continued as before, with manual labor each day combined with certain hours of study. The students did not want to give up the present opportunity of learning how to labor and how to study. If this is their choice under the most disadvantageous circumstances, what influence will it have when the school buildings are up and there are more favorable surroundings for the students?

The building they now occupy, the only one at all fit for the purpose, was an old hotel which we rented and are using to its fullest capacity. Four tents pitched in an adjoining paddock are also occupied by students. Every morning at six o'clock the members of the school are called together for morning worship and Bible study. These occasions have proved a blessing....

I spoke to the students eight mornings. The Lord Jesus was indeed in our assembly. The congregation averaged from twenty-six to thirty. In the first meetings the spirit of intercession came upon me, and all were sensible that the Lord heard our prayers. Then I spoke about thirty minutes, and the Lord gave me words for those assembled. These seasons were most profitable; the testimonies of the students following gave evidence that the Holy Spirit was giving to all glimpses of the things of God.

The spiritual impressions became more marked as the meetings progressed. The divine presence was with us. The sympathies and sentiments of those present became inspired with power and favor. Hearts were susceptible to the influence of the Holy Spirit, and decided changes were wrought in minds and character. The Spirit of God was working upon human agents. I praise the Lord for the encouraging influence of His Spirit upon my own heart. We all felt that the Lord was cooperating with us to lead us to will, to resolve, and act.

The Lord does not propose to perform for us either the willing or the doing. This is our proper work. As soon as we earnestly enter upon the work, God's grace is given to work in us to will and to do, but never as a substitute for our effort. Our souls are to be aroused to cooperate. The Holy Spirit works the human agent, to work out our own salvation. This is the practical lesson the Holy

Hope in the Soil

Spirit is striving to teach us. For it is God which worketh in you both to will and to do of His good pleasure.

I never had a deeper sense of the precious truth and its power upon human minds than when addressing those students in the early meetings. Morning after morning I felt charged with a message from God. I also had special freedom in speaking twice upon the Sabbath. At every meeting several unbelievers were present, and they were much affected as the truth was presented. If we had a suitable place for meeting we could invite the neighbors to come in. But our long, narrow dining room crowded as closely as if packed is not a very suitable place for worship. I am assigned a little space in the corner of the room, and am packed up close to the wall. Nevertheless the Lord Jesus is in the assembly. We know it. Some souls are thinking very seriously now upon the subject of the truth.

We all know that the most severe and intense soul struggles belong to the hour of the great resolve to act out the convictions upon the human heart. The consecration of the soul to God is committing the keeping of the soul to One who has purchased its freedom at an infinite price, and then we are to follow on to know the Lord, that we may know His goings forth are prepared as the morning. To obey is better than sacrifice. The whole work of the Christian is comprised in willing and doing.

The students work hard and faithfully. They are gaining in strength of nerve and in solidity as well as activity of muscles. This is the proper education which will bring forth from our schools young men who are not weak and inefficient, who have not a one-sided education, but an all-round physical, mental, and moral training. The builders of character must not forget to lay the foundation which will make education of the greatest value. This will require self-sacrifice, but it must be done. The physical training will, if properly conducted, prepare for mental taxation. But the one alone always makes a deficient man. The physical taxation combined with mental effort keeps the mind and morals in a more healthful condition, and far better work is done. Under this training students will come forth from our schools educated for practical life, able to put their intellectual capabilities to the best use. Physical and mental exercise must be combined if we would do justice to our students. We have been working on this plan here with complete satisfaction, notwithstanding the inconvenience under which students have to labor.

I came here and began work on my place so earnestly that it inspired all with fresh zeal, and they have been working with a will, rejoicing that they have the privilege. We have provoked one another to zeal and good works. The school workers were afraid I would plant the first trees, and now both they and and I have the satisfaction of having the first genuine orchards in this vicinity. Some of our trees will yield fruit next year, and the peaches will bear quite a crop in two years.

Mr.-----, from whom we bought our trees, lives about twenty miles from here. He has an extensive and beautiful orchard. He says that we have splendid fruit land.

Well, the school has made an excellent beginning. The students are learning how to plant trees, strawberries, etc.; how they must keep every sprangle and fiber of the roots uncramped in order to give them a chance to grow. Is not this a most precious lesson as to how to treat the human mind, and the body as well—not to cramp any of the organs of the body, but give them ample room to do their work? The mind must be called out, its energies taxed. We want men and women who can be energized by the Spirit of God to do a complete work under the Spirit's guidance. But these minds must be cultivated, employed, not lazy and dwarfed by inaction. Just so men and women and children are wanted who will work the land, and use their tact and skill, not with a feeling that they are menials, but that they are doing just such noble work as God gave to Adam and Eve in Eden, who loved to see the miracles wrought by the divine Husbandman. The human agent plants the seed, and God waters it and causes His sun to shine upon it, and up springs the tiny blade. Here is the lesson God gives to us concerning the resurrection of the body, and the renewing of the heart. We are to learn of spiritual things from the development of the earthly.

We are not to be put about and discouraged about temporal things because of apparent failures, nor should we be disheartened by delay. We should work the soil cheerfully, hopefully, gratefully, believing that the earth holds in her bosom rich stores for the faithful worker to garner, richer than gold or silver. The niggardliness laid to her charge is false witness. With proper, intelligent cultivation the earth will yield its treasures for the benefit of man.

The spiritual lessons to be learned are of no mean order. The seeds of truth sown in the soil of the heart will not all be lost, but will spring up, first the blade, then the ear, and then the corn in the ear. God said in the beginning, Let the earth bring forth grass, the herb yielding seed, and the fruit tree yielding fruit. God created the seed as He did the earth, by the divine word. We are to exercise our reasoning powers in the cultivation of the earth, and to have faith in the word of God that has created the fruit of the earth for the service of man.

The cultivation of our lands requires the exercise of all the brainpower and tact we possess. The lands around us testify to the indolence of men. We hope to arouse to action the dormant senses. We hope to see intelligent farmers, who will be rewarded for their earnest labor. The hand and heart must cooperate, bringing new and sensible plans into operation in the cultivation of the soil. We have here seen the giant trees felled and uprooted, we have seen the plowshare pressed into the earth, turning deep furrows for the planting of young trees and the sowing of the seed. The students are learning what plowing means, and that the hoe and the shovel, the rake and the harrow, are all implements of honorable and profitable industry. Mistakes will often be made, but error lies close beside truth. Wisdom will be learned by failures, and the energy that will make a beginning gives hope of success in the end. Hesitation will keep things back, precipitancy will alike retard, but all will serve as lessons if the human agents will have it so.

In the school that is started here in Cooranbong, we look to see real success in agricultural lines, combined with a study of the sciences. We mean for this place to be a center, from which shall irradiate light, precious advanced knowledge that shall result in the working of unimproved lands, so that hills and valleys shall blossom like the rose. For both children and men, labor combined with mental taxation will give the right kind of all-round education. The cultivation of the mind will bring tact and fresh incentives to the cultivation of the soil.

There will be a new presentation of men as breadwinners, possessing educated, trained ability to work the soil to advantage. Their minds will not be overtaxed and strained to the uttermost with the study of the sciences. Such men will break down the foolish sentiments that have prevailed in regard to manual labor. An influence will go forth, not in loud-voiced oratory, but in real inculcation of ideas. We shall see farmers who are not coarse and rough and slack, careless of their apparel and of the appearance of their homes; but they will bring taste into farmhouses. Rooms will be sunny and inviting. We shall not see blackened ceilings, covered with cloth full of dust and dirt. Science, genius, intelligence, will be manifest in the home. The cultivation of the soil will be regarded as elevating and ennobling. Pure, practical religion will be manifested in treating the earth as God's treasure-house. The more intelligent a man becomes, the more should religious influence be radiating from him. And the Lord would have us treat the earth as a precious treasure, lent us in trust.

Ellen G. White

Chapter #7: No Time to Lose

BECAUSE WE LIVE IN SUCH PERILOUS TIMES, THE PEOPLE OF GOD SHOULD MOVE QUICKLY TO LOCATE IN THE COUNTRY AND LEARN TO RAISE THEIR OWN FOOD: When the power invested in kings is allied to goodness, it is because the one in responsibility is under the divine dictation. When power is allied with wickedness, it is allied to Satanic agencies, and it will work to destroy those who are the Lord's property. The Protestant world have set up an idol sabbath in the place where God's Sabbath should be, and they are treading in the footsteps of the Papacy. For this reason I see the necessity of the people of God moving out of the cities into retired country [places,] where they may cultivate the land and raise their own produce. Thus they may bring their children up with simple, healthful habits. I see the necessity of making haste to get all things ready for the crisis. (Letter 90, 1897)

PARENTS SHOULD MOVE FROM THE CITIES AS QUICKLY AS POSSIBLE: To parents who are living in the cities the Lord is sending the warning cry, Gather your children into your own houses; gather them away from those who are disregarding the commandments of God, who are teaching and practicing evil. Get out of the cities as fast as possible. (Medical Ministry p. 310)

WE MUST DO OUR PART TO SEPARATE FROM EVIL INFLUENCES: If we place ourselves under objectionable influences, can we expect God to work a miracle to undo the results of our wrong course? No, indeed. Get out of the cities as soon as possible, and purchase a little piece of land where you can have a garden, where your children can watch the flowers growing, and learn from them lessons of simplicity and purity. Consider the lilies of the field, how they grow; they toil not, neither do they spin: And yet I say unto you, That even Solomon in all his glory was not arrayed like one of these (Matthew 6:28, 29). Parents, point your children to the beautiful things of God's creation, and from these things teach them of His love for them. Point them to the lovely flowers—the roses and the lilies and the pinks—and then point them to the living God. (Manuscript Release 12 p. 31)

WORK WITH PERSEVERING ENERGY: Much depends upon laying our plans according to the word of the Lord and with persevering energy carrying them out. (Testimonies to the Church Vol. 6 p. 178)

Hope in the Soil

THE IMPORTANCE OF FOLLOWING GOD'S PLAN: Not by one word, not by many words, but by every word that God has spoken, shall man live. We cannot disregard one word, however trifling it may seem to us, and be safe. (Mount of Blessing p. 52)

OBEDIENCE A CONDITION TO SUCCESS: Obedience to every word of God is another condition of success. Victories are not gained by ceremonies or display, but by simple obedience to the highest General, the Lord God of heaven. He who trusts in this Leader will never know defeat. Defeat comes in depending on human methods, human inventions, and placing the divine secondary. (Testimonies to the Church Vol. 6 p. 140)

SOME BENEFITS FROM FOLLOWING GOD'S PLAN: The more closely His plan of life is followed, the more wonderfully will He work to restore suffering humanity. The sick need to be brought into close touch with nature. An outdoor life amid natural surroundings would work wonders for many a helpless and hopeless invalid. (Ministry of Healing 262)

EDUCATIONAL LESSONS SHOULD BE CHOSEN WITH THE HARD, UNCONGENIAL LABOR AHEAD IN MIND: All who become co-workers with Christ will have a great deal of hard, uncongenial labor to perform, and their lessons of instruction should be wisely chosen, and adapted to their peculiarities of character, and the work which they are to pursue. (Gospel Workers, 1892 317)

BIBLE RELIGION TO BE WRAPPED UP IN ALL THAT WE DO OR SAY— INCLUDING FARMING: Bible religion is to be interwoven with all we do or say.... They are to be united in all human pursuits, in mechanical and agricultural labors, in mercantile and scientific enterprises.... It is just as essential to do the will of God when erecting a building as when taking part in a religious service.... (My Life Today p. 117)

ELLEN WHITE PUT PLANTING HER FRUIT TREES EVEN BEFORE BUILDING HER HOUSE: I determined to set my trees, even before the foundation of the house was built. We broke up only furrows, leaving large spaces unplowed. Here in these furrows we planted our trees the last of September, and lo, this year they were loaded with beautiful blossoms and the trees were loaded with fruit. It was thought best to pick off the fruit, although the trees had obtained a growth that seemed almost incredible. The small amount of fruit—peaches and nectarines—have served me these three weeks. They were delicious, early peaches. We have later peaches—only a few left to mature as samples. Our pomegranates looked beautiful in full bloom. Apricots were trimmed back in April and June, but they threw up their branches and in five weeks, by measurement, had a thrifty growth of five and eight feet. (Manuscript Release 8 p. 252)

NO TIME TO ENGAGE IN SPECULATION OR HAPHAZARD MOVEMENTS: True education means much. We have no time now to spend in speculative ideas, or in haphazard movements. The evidences that the coming of Christ is near are many and are very plain, and yet many who profess to be looking for Him are asleep. We are not half as earnest as we ought to be to gather up the important truths that are for our admonition, upon whom the ends of the world are come. Unless we understand the importance of passing events, and make ready to stand in the great day of God, we shall be registered in the books of heaven as unfaithful stewards. The watchman is to know the time of the night. Everything is now clothed with a solemnity that all who believe the truth should feel and understand. They should act in reference to the great day of God. (Manuscript Release 8 p. 154)

OUR SCHOOLS NEED NOT REMAIN HELPLESS, IN STATES OF UNCERTAINTY: Opposing circumstances will and should create a firm determination to overcome them. One barrier broken down will give greater ability and courage to go forward. Fate has not woven its meshes about the workings of our schools that they need to remain helpless and in uncertainty. Press in the right direction, and make a change, solidly, intelligently. Then circumstances will be your helpers and not your hindrances. (Manuscript Release 8 p. 200)

REGARDLESS OF MISTAKES OF THE PAST, EARNEST EFFORTS TOWARDS RIGHT TO BE MADE: So much work of a faulty nature has been done in the school at Huntsville that it will now require stern efforts to restore the work to healthfulness, but such efforts should be put forth. Many discouragements have come in, but the Lord will let His blessing rest upon those who will take hold of the work thoroughly and in earnest. There is a special need of intense earnestness. Take hold with heart and mind and strength to redeem the farm, that it may be, as it has been presented to me, a beautiful place, well pleasing to the Lord, a spectacle to angels and to men. (A Place Called Oakwood p. 119)

THE TIME WILL COME WHEN GOD'S PEOPLE WILL MOVE AWAY FROM THE CITIES AND LIVE IN SMALL COMPANIES BY THEMSELVES: The plagues of the last days are to be poured out on the inhabitants of the world who have shown marked contempt for the law of God. God's people should seek to reach the people of the world, proclaiming the truth as it is found in His Word. But the time will come when they will have to move away from the cities, and live in small companies, by themselves. (Manuscript Release 17 p. 350)

THE WORK HAS BEEN HINDERED BY FAILURE TO FOLLOW GOD'S EDUCATIONAL PLAN: I have been shown that in our educational work we are not to follow the methods that have been adopted in our older established schools. There is among us too much clinging to old customs, and because of this we are far behind where we should be in the development of the third angel's message. Because men could not comprehend the purpose of God in the plans laid before us for the education of workers, methods have been followed in some of our schools which have retarded rather than advanced the work of God. Years have passed into eternity with small results, that might have shown the accomplishment of a great work. If the Lord's will had been done by the workers in earth as the angels do it in heaven, much that now remains to be done would be already accomplished, and noble results would be seen as the fruit of missionary effort. (Counsels to Parents, Teachers and Students p. 533)

TRUE EDUCATION MUST HAPPEN BEFORE WE CAN BREAK EVERY YOKE: Before we can carry the message of present truth in all its fullness to other countries, we must first break every yoke. We must come into the line of true education, walking in the wisdom of God, and not in the wisdom of the world. God calls for messengers who will be true reformers. We must educate, educate, to prepare a people who will understand the message, and then give the message to the world. (Special Testimonies Series B p. 30)

Chapter #7: No Time to Lose

(1): Promises of Help and Success

And all thy children shall be taught of the Lord; and great shall be the peace of thy children. (Isaiah 54:13)

To him that overcometh will I give to eat of the tree of life, which is in the midst of the paradise of God. Revelation 2:7

GOD WILL BLESS FARM MANAGERS AND TEACHERS WHO WORK WITH THE HOLY SPIRIT, AND MAKE THEM WISE: If the managers of this farm and the teachers in the school will receive the Holy Spirit to work with them, they will have wisdom in their management, and God will bless their labors. (Australasian Union Conference Record, July 31, p. 1899)

If you walk in my statutes and observe my commandments and do them, then I will give you your rains in their season, and the land shall yield its increase, and the trees of the field shall yield their fruit. (Lev. 26:3-4)

And the Lord will make you abound in prosperity. The Lord will open to you His good treasury the heavens, to give the rain of your land in its season and to bless the work of your hands. And the Lord will make you the head, and not the tail; and you shall tend upward only, and not downward; if you obey the commandments of the Lord your God, which I command you this day, being careful to do them, and if you do not turn aside from any of the words which I command you this day, to the right hand or to the left. (Deut. 28:11-14)

Believe in the Lord your God, so shall ye be established; believe His prophets, so shall ye prosper. (2 Chron. 20:20)

GOD WILL HELP HIS PEOPLE FIND HOMES OUTSIDE OF THE CITIES: Parents can secure small homes in the country, with land for cultivation where they can have orchards and where they can raise vegetables and small fruits to take the place of flesh-meat, which is so corrupting to the lifeblood coursing through the veins. On such places the children will not be surrounded with the corrupting influences of city life. God will help His people to find such homes outside of the cities. (Medical Ministry p. 310)

MISTAKES WILL BE MADE—BUT ERROR LIES CLOSE BESIDE TRUTH: Mistakes will often be made, but every error lies close beside truth. Wisdom will be learned by failures, and the energy that will make a beginning gives hope of success in the end. Hesitation will keep things back, precipitancy will alike retard, but all will serve as lessons if the human agents will have it so. (Letter 47a, 1895)

GOD PROSPERS SCHOOLS CONDUCTED ACCORDING TO HIS DESIGN: God will bless the work of those schools that are conducted according to His design. When we were laboring to establish the educational work in Australia, the Lord revealed to us that this school must not pattern after any schools that had been established in the past. This was to be a sample school. The school was organized on the plan that God had given us, and He has prospered its work. (Series B, No. 11, pp. 338, 339)

EXPECT THE BLESSING OF GOD: …there must be land for orchards and gardens, that students may have physical exercise combined with mental taxation, and half and some wholly pay their way at school…Education must be given in regard to tilling the soil, and we must expect that the Lord will bless this effort. (Welfare Ministry p. 184-185)

CULTIVATED LAND WILL YIELD ITS TREASURES: If the land is cultivated, it will, with the blessing of God, supply our necessities. With proper intelligent cultivation the earth will yield its treasures for the benefit of man. The mountains and hills are changing; the earth is waxing old like a garment; but the blessing of God, which spreads a table for His people in the wilderness, will never cease. (Testimonies to the Church Vol. 6 p. 178)

ATTENTION CALLED TO THE LAST GREAT WORK TO SAVE THE PERISHING—THROUGH OUR SCHOOLS: Our work is reformatory, and it is the purpose of God that through the excellence of the work done in our educational institutions the attention of the people shall be called to the last great effort to save the perishing. (Testimonies to the Church Vol. 6 p. 126)

GOD'S GREAT COVENANT: God's great covenant declares that 'while the earth remaineth, seedtime and harvest shall not cease.' Genesis 8:22. In the confidence of this promise the husbandman tills and sows. (Education p. 105)

GOD WILL TEACH FARMERS: Does he who plows for sowing plow continually? Does he continually open and harrow his ground? When he has leveled its surface, does he not scatter dill, sow cumin, and put in wheat in rows and barley in its proper place, and spelt (a variety of wheat) as the border? For he is instructed right; his God teaches him. This comes from the Lord of hosts; he is wonderful in counsel, and excellent in wisdom. (Isa. 28:24-29, RSV)

WISDOM TO BE LEARNED THROUGH FAILURES: Wisdom will be learned by failures, and the energy that will make a beginning gives hope of success in the end. (Testimonies to Ministers and Gospel Workers p. 244)

TAKE COURAGE: Take courage, all you people of the land, says the Lord; work, for I am with you, says the Lord of hosts. (Haggai 2:4)

MORE WORDS OF ENCOURAGEMENT: Opposing circumstances should create a firm determination to overcome them. One barrier broken down will give greater ability and courage to go forward. Press in the right direction, and make a change, solidly, intelligently. Then circumstances will be your helpers and not your hindrances. Make a beginning. The oak is in the acorn. (Testimonies to the Church Vol. 6 p. 145)

Whatever is to be done at His command may be accomplished in His strength. All His biddings are enablings. (Christ's Object Lessons p. 333)

Obedience to every word of God is another condition of success. Victories are not gained by ceremonies or display, but by simple obedience to the highest General, the Lord God of heaven. He who trusts in this Leader will never know defeat. Defeat comes in depending on human methods, human inventions, and placing the divine secondary. Obedience was the lesson that the Captain of the Lord's host sought to teach the vast armies of Israel—obedience in things in which they could see no success. When there is obedience to the voice of our Leader, Christ will conduct His battles in ways that will surprise the greatest powers of earth. (Testimonies to the Church Vol. 6 p. 140)

Fear thou not; for I am with thee: be not dismayed; for I am thy God: I will strengthen thee; yea, I will help thee; yea, I will uphold thee with the right hand of my righteousness. (Isa. 41:10)

(2): Results of Failure to Obey

DISASTER WILL COME UPON THOSE WHO REMAIN IN THE CITIES: If our people regard God's instruction as of value, they will move away from the city, so that they will not be pained by its revolting sights, and that their children will not be corrupted by its vices. Those who choose to remain in the cities, surrounded by the houses of unbelievers, must share the disaster that will come upon them. (Manuscript Release 17 p. 350)

HALF-CONVERTED TEACHERS AND MANAGERS HOLD BACK THE WORK: To some extent the Bible has been introduced into our schools, and some efforts have been made in the direction of reform; but it is most difficult to adopt right principles after having been so long accustomed to popular methods. The first attempts to change the old customs brought severe trials upon those who would walk in the way which God has pointed out. Mistakes have been made, and great loss has resulted. There have been hindrances which have tended to keep us in common, worldly lines, and to prevent us from grasping true educational principles. To the unconverted, who view matters from the lowlands of human selfishness, unbelief, and indifference, right principles and methods have appeared wrong. Some teachers and managers who are only half converted are stumbling blocks to others. They concede some things and make half reforms; but when greater knowledge comes, they refuse to advance, preferring to work according to their own ideas. In doing this they pluck and eat of that tree of knowledge which places the human above the divine. Now therefore fear the Lord, and serve Him in sincerity and in truth: and put away the gods which your fathers served on the other side of the flood, and in Egypt; and serve ye the Lord. And if it seem evil unto you to serve the Lord, choose you this day whom ye will serve. If the Lord be God, follow Him: but if Baal, then follow him. (Joshua 24:14-15, 1 Kings 18:21). We should have been far in advance of our present spiritual condition had we moved forward as the light came to us. (Testimonies for the Church, Volume 6, page 141)

DANGERS OF CLINGING TO CUSTOMS: There is a constant danger that our educators will travel over the same ground as did the Jews, conforming to customs, practices, and traditions which God has not given. With tenacity and firmness some cling to old habits and a love of various studies which are not essential, as if their salvation depended upon these things. In doing this they turn away from the special work of God and give to the students a deficient, a wrong education. (Testimonies to the Church Vol. 6 p. 150)

That which the Lord has spoken concerning the instruction to be given in our schools is to be strictly regarded. (Testimonies to the Church Vol. 6 p. 142)

Much depends upon laying plans according to the word of the Lord and with persevering energy carrying them out. (Testimonies to the Church Vol. 6 p. 178)

Made in the USA
Las Vegas, NV
02 May 2021

22360787R00111